Differential Calculus and its Applications

(As per NEP-2020 Syllabus)

EKTA N. JAYSWAL
Assistant Professor
Department of Mathematics
Institute of Technology
Nirma University
Ahmedabad

NITA H. SHAH
Emeritus Professor
Department of Mathematics &
School of Emerging Science and Technology
Gujarat University
Ahmedabad

PHI Learning Private Limited
Delhi-110092
2025

*In fond memory of **Shri Asoke K. Ghosh** (October 1942 – February 2024), Founder Chairman and Managing Director of PHI Learning, whose vision endlessly inspires.*

The Legacy Continues....

Published by Pushpita Ghosh, PHI Learning Private Limited, Rimjhim House, 111, Patparganj Industrial Estate, Delhi-110092 and Printed by Syndicate Binders, A-20, Hosiery Complex, Noida, Phase-II Extension, Noida-201305 (N.C.R. Delhi).

₹ 450.00

DIFFERENTIAL CALCULUS AND ITS APPLICATIONS
Ekta N. Jayswal and Nita H. Shah

ISBN-978-93-5443-980-3 (Print Book)
ISBN-978-93-5443-790-8 (e-Book)

The export rights of this book are vested solely with the publisher.

Contents

Preface

Differential Calculus is one of the foundational pillars of mathematics, serving as both a powerful analytical tool and a gateway to advanced studies in science, engineering, and economics. This book, *Differential Calculus and its Applications* has been designed for students, educators, and professionals seeking a deeper understanding of calculus and its myriad applications.

The journey into Differential Calculus begins with its conceptual core: the derivative. The derivative is not merely a mathematical abstraction; it represents change, growth, and rates—phenomena that permeate every aspect of our lives. From understanding the velocity of a moving car to optimizing industrial processes and modeling economic behavior, derivatives provide insight into the dynamic nature of the world around us.

This book is structured to provide both theoretical rigor and practical insights. The initial chapters establish a solid foundation by introducing fundamental concepts such as limits, continuity, and differentiation. These concepts are illustrated with clarity and precision through numerous examples and exercises designed to reinforce understanding.

Subsequent chapters focus on Differential Calculus applications. Readers will explore techniques for solving real-world problems, including optimization, curve sketching, and motion analysis. Each application has step-by-step explanations, bridging the gap between abstract mathematical theory and practical utility.

This book also delves into modern applications of Differential Calculus, highlighting its role in data science, artificial intelligence, and complex systems analysis. These sections aim to inspire readers by showcasing the relevance of calculus in cutting-edge technologies and interdisciplinary research.

We are sure that students and the teaching community will find this new book useful and easy to understand. Any suggestion for improving the book will be highly appreciated and gratefully acknowledged.

We wish to thank the Publisher, PHI Learning, for the careful processing of the manuscript, both at the editorial and production levels.

Ekta N. Jayswal
Nita H. Shah

1 Function, Limit, and Continuity

Before entering into the concept of function, as we know about sets and the Cartesian product of two sets, we introduce the term relation.

1.1 RELATION

DEFINITION: Let A and B two non-empty sets. A Relation R from sets A to B is a subset of the Cartesian product set $A \times B$.

The subset is derived from this relation R and contains ordered pairs in which the first element is from the set A and the second element is from the set B. The set A is called the domain of the relation R and the set B is called the range of the relation R.

For example, $R = \left\{(2, -4), \left(3, \frac{1}{3}\right), (4, 1)\right\}$ is a relation; then the domain of R = {2, 3, 4} and range of $R = \left\{-4, \frac{1}{3}, 1\right\}$.

There exist important general properties for relation R. Let R be a relation on a set A.

1. R is **reflexive** if and only if each element is related to itself, *i.e.*, if $(x, x) \in R \; \forall x \in A$.
2. R is **symmetric** if and only if any one element is related to the second element then the second element should relate to the first, *i.e.*, if $(x, y) \in R$ then $(y, x) \in R$.
3. R is **transitive** if and only if one element is related to a second element then that second element is related to a third then the first element is related to the third, *i.e.*, if $(x, y) \in R$ and $(y, z) \in R$ then $(x, z) \in R$.

Function, limit, and continuity are the fundamental study of calculus. In this chapter, we will discuss the definition, interpretation, and examples related to these concepts and their real-life applications.

1.2 FUNCTIONS

The function is a relationship between an input set having a unique output. It can be signified by a graph, an equation, an arithmetical table, etc. A function is a mathematical tool that defines the factual world in mathematical terms.

As function is related to input-output, one can understand by example that if we input the radius of a circle then we get the area of a circle, a thermometer, weekly/monthly salary depending upon the worked hourly payment rate, etc.

In every situation, a variable quantity (say y), depends upon another variable quantity (say x). Symbolically one can write as, $y = f(x)$, can read as "y equals to f of x" and also say that "y is a function of x". f represents the function; x (input) is the independent variable and y (output) is the dependent variable.

DEFINITION: A function f from set X to a set Y is a rule that assigns a unique element $f(x) \in Y$ to each element $x \in X$. Notation $f : X \rightarrow Y$ (Read as: f is from X to Y).

The set X of all probable input values is called the **domain** of the function and the set of all values $f(x)$ is called the **range** of the function.

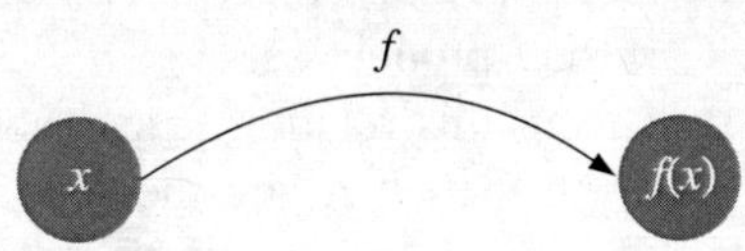

FIGURE 1.1 A diagram showing a function.

Let's discuss some functions:

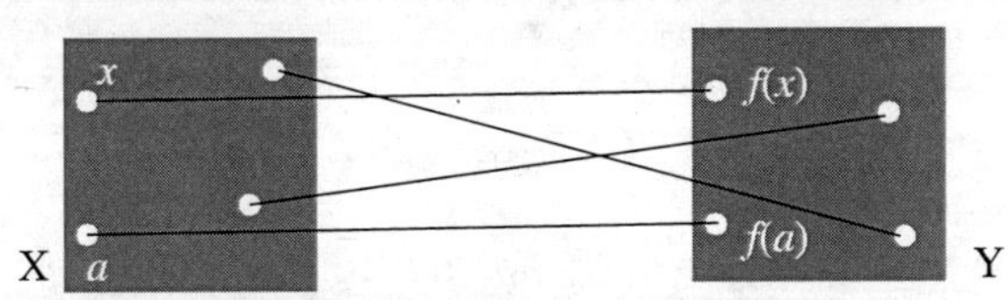

(a) A function $f : X \rightarrow Y$

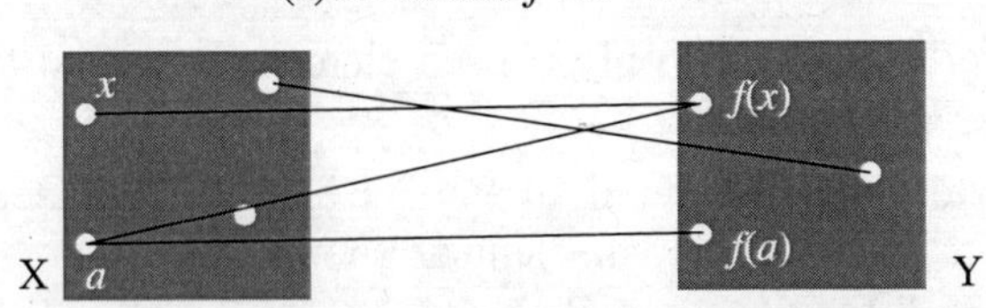

(b) This is not a function as "a" associated with two different values of Y

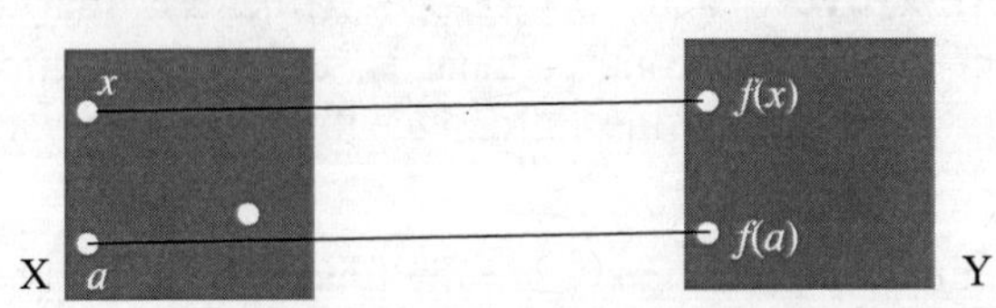

(c) It is not a function as one member of X not associated with any value of Y

FIGURE 1.2 *(Contd.)*

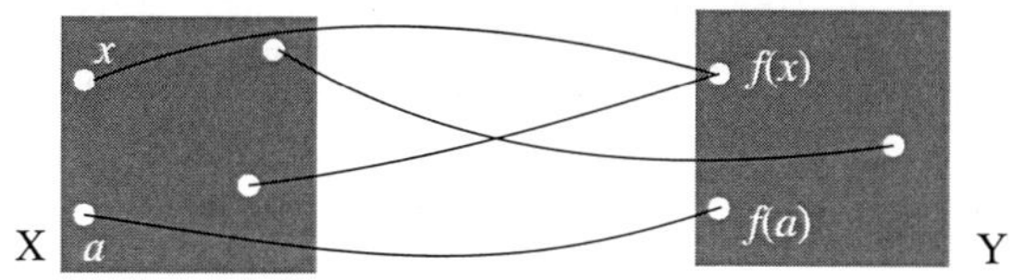

(d) This is a function as every value of X is connected with a unique member of Y

FIGURE 1.2 Different types of functions.

EXAMPLE 1.1 Evaluate some examples of domain and range.

1. $y = x^2$:

   Ans: Here, the domain (x) will take any real value and the range (y) will be a real non-negative number. So, the domain $(-\infty, \infty)$ and range will be $[0, \infty)$.

2. $y = \sqrt{x}$:

   Ans: Now, in this example range (y) only takes real value if the domain $x \geq 0$. So, domain and range are $[0, \infty)$ respectively.

3. $z = \sqrt{(a-x)(x-b)} + \sqrt{(c-y)(y-d)}$, where, $a < b$; $c < d$

   Ans: This $(a - x)(x - b)$ is non-negative if $a \leq x \leq b$ and $(c - y)(y - d)$ is non-negative if $c \leq y \leq d$. Now, the value of z is a function of (x, y) defined for the domain (rectangular domain bounded by lines $x = a$, $x = b$; $y = c$, $y = d$).

Graph of Function

Let $f : X \rightarrow Y$ be the function. Then the graph of the function f consists of the points of the Cartesian plane and shows the certain value of $y \in Y$ corresponding to each value of $x \in X$ defined.

EXAMPLE 1.2 $y = x^2$ in the interval $[-3, 3]$.

Ans: First evaluate the values of the input-output xy-pairs satisfying equation $y = x^2$. We can write one table format and try to plot the graph as a smooth curve.

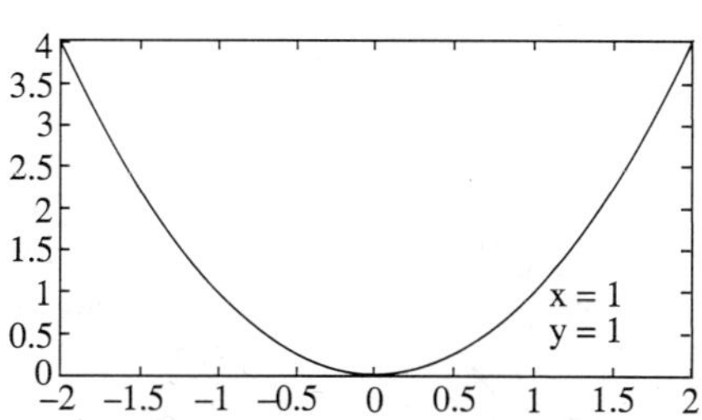

FIGURE 1.3 Graph of $y = x^2$.

x	$y = x^2$
–3	9
–2	4
–1	1
0	0
1	1
2	4
3	9

About Standard Functions

1. **Constant Function:** If a function $f : R \to R$ defined as $f(x) = c$, $\forall x$ where, c is a constant value then f is called a constant function. Domain = R, Range = $\{c\}$.

   For example, $f(x) = 2$.

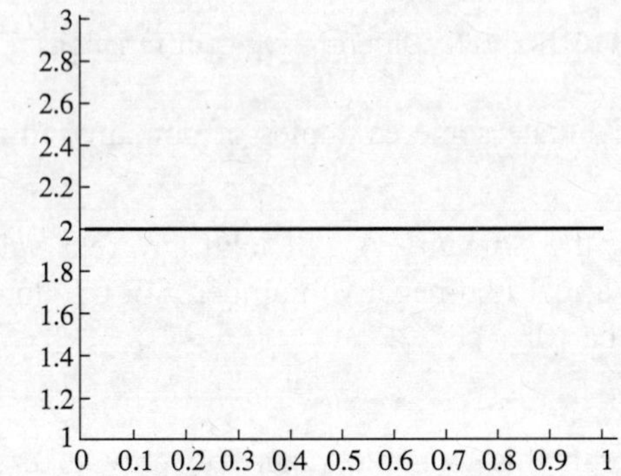

FIGURE 1.4 Constant function graph.

2. **Piecewise-Defined Function:** A function is defined by different parts with different formulas in its domain. For example, the **modulus function** $f : R \to R$ defined as $f(x) = |x| = \begin{cases} x & x \geq 0 \\ -x & x < 0 \end{cases}$. Domain = R and Range = $R^+ \cup \{0\}$.

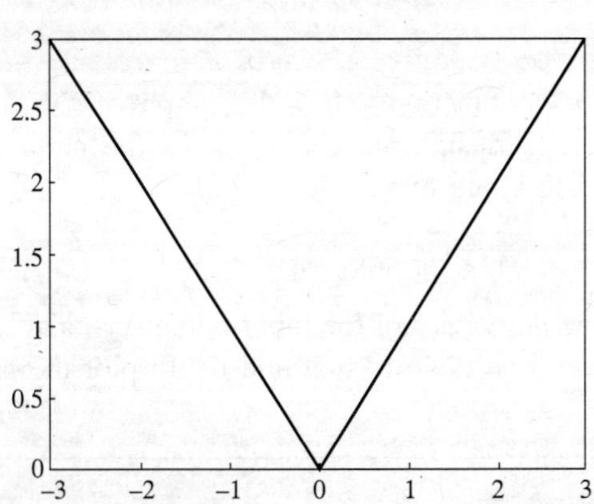

FIGURE 1.5 Modulus function graph.

3. **Greatest Integer Function:** The function $f : R \to R$ at any number x is defined by $f(x) = [x]$, $x \in R$. It assumes the greatest integer less than or equal to that value x, which is called the **greatest integer function**.

   For example, $f(x) = [x] = -1$ for $-1 \leq x < 0$

   $f(x) = [x] = 0$ for $0 \leq x < 1$

   $f(x) = [x] = 1$ for $1 \leq x < 2$ and so on

4. **Identity Function:** The function $f : R \to R$ is defined by $f(x) = x, \forall x$ then this function is called an **Identity function.** Domain = R and Range = R.

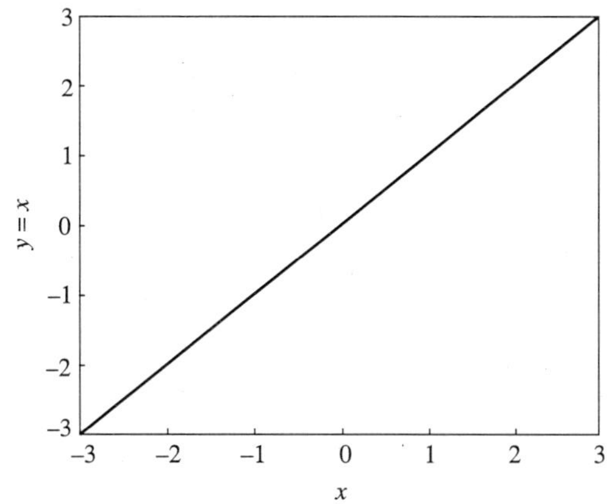

FIGURE 1.6 Identity function graph.

5. **Monotonic Functions:** Let $f : R \to R$ be a function defined on an interval I and let x_1 and x_2 be two points in I.
   (a) If $f(x_1) < f(x_2)$ whenever $x_1 < x_2$ then, f is called **increasing** function on I.
   (b) If $f(x_1) > f(x_2)$ whenever $x_1 < x_2$ then, f is called **decreasing** function on I.

For example, $\sin x, x \in \left[0, \frac{5\pi}{2}\right]$ the function is increasing on $\left[0, \frac{\pi}{2}\right]$ and decreases on $\left[\frac{\pi}{2}, \frac{3\pi}{2}\right]$, and so on.

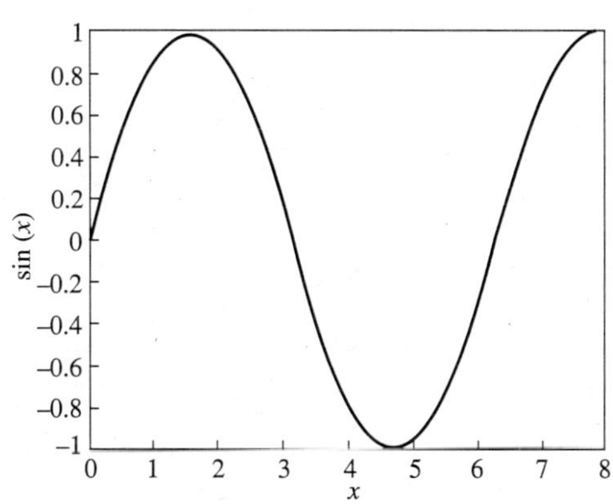

FIGURE 1.7 Sine function graph.

6. **Polynomial Function:** A function $f : R \to R$ defined by $f(x) = a_n x^n + a_{n-1}x^{n-1} + ... + a_1 x + a_0$, $n \in N$ and $a_0, a_1, ..., a_n \in R$ having a domain $(-\infty, \infty)$ is called a **polynomial**. For example, $f(x) = 8x^4 - 14x^2 - 9x^2 + 11x - 1$.

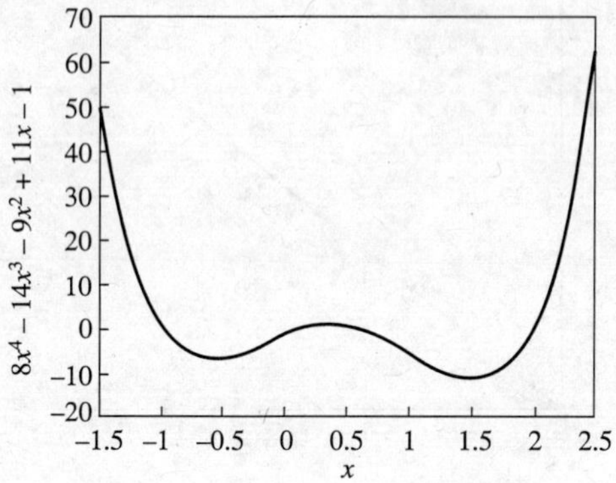

FIGURE 1.8 Polynomial function graph.

7. **Logarithmic and Exponential Function:** The function of the form $f(x) = \log_a x$, where the base $a \neq 1$ is a positive constant known as the **logarithmic** function and $f(x) = a^x$ where $a \neq 1$ is a positive constant called the **exponential** function.

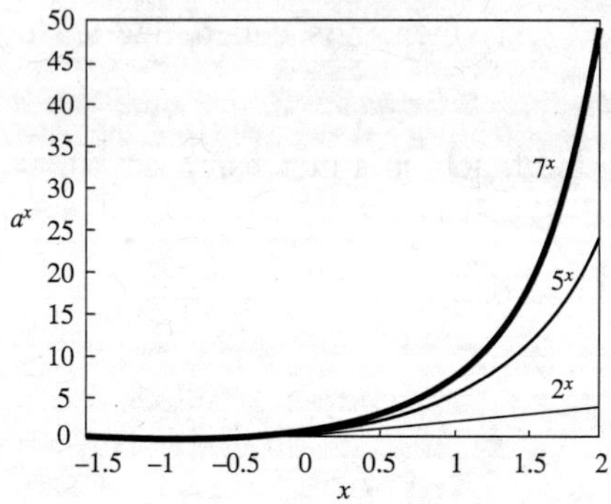

FIGURE 1.9 Exponential function graph.

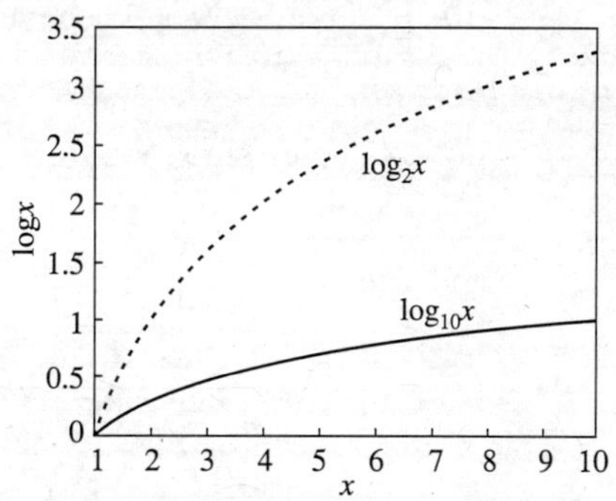

FIGURE 1.10 Logarithmic function graph.

1.3 OPERATIONS ON REAL-VALUED FUNCTIONS

Functions can be added, subtracted, and multiplied like real numbers to produce new real-valued functions. Let $f : R \to R$ and $g : R \to R$ be two real functions and their domains defined by $D(f)$ and $D(g)$ then $\forall x \in D(f) \cap D(g)$ we define,

$$(f + g)(x) = f(x) + g(x)$$

$$(f - g)(x) = f(x) - g(x)$$

$$(fg)(x) = f(x)\, g(x)$$

Functions also multiplied by a scalar (say $\alpha \in R$), then $(\alpha f)(x) = \alpha f(x)$.

Functions can be divided (when $g(x) \neq 0$) and then f/g defined by for $\forall x \in D(f) \cap D(g)$,

$$\left(\frac{f}{g}\right)(x) = \frac{f(x)}{g(x)}$$

EXAMPLE 1.3 Let $f(x) = x + 2$ and $g(x) = x^2 - 4$ be two functions. The domains are $D(f) = D(g) = (-\infty, \infty)$, for division $D(f) = (-\infty, \infty)$, $D(g) = (-\infty, \infty) - \{-2, 2\}$. Now we have,

$$(f+g)(x) = x + 2 + x^2 - 4 = x^2 + x - 2$$

$$(f-g)(x) = x + 2 - x^2 + 4 = -x^2 + x + 6$$

$$(f.g)(x) = (x+2)(x^2 - 4) = x^3 + 2x^2 - 4x - 8$$

$$(3f)(x) = 3(x+2) = 3x + 6$$

$$(f/g)\,(x) = (x+2)/(x^2 - 4) = 1/(x-2)$$

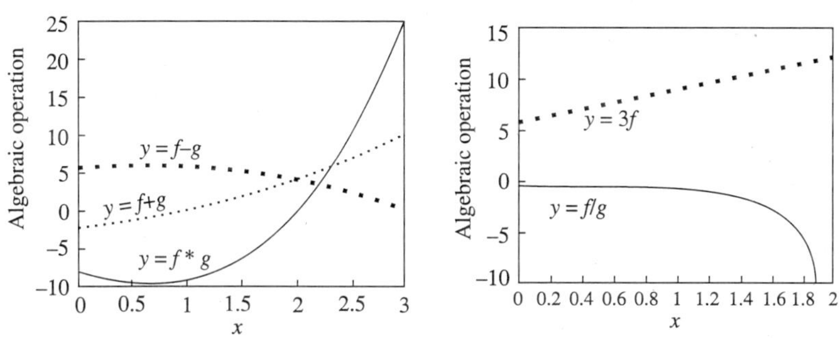

FIGURE 1.11 Algebraic operation graph.

Other than algebraic operations, the concept of the composition of two functions is important. Composition is to combine two functions. If $f : A \rightarrow B$ and $g : C \rightarrow D$ be two functions with range $(f(x)) \subset$ domain $(g(x))$, then the **Composite function** $h = g \circ f = g(f(x)) : A \rightarrow C$ is defined. In this scenario, we first operate $f(x)$, and then we apply $g(x)$ onto that function.

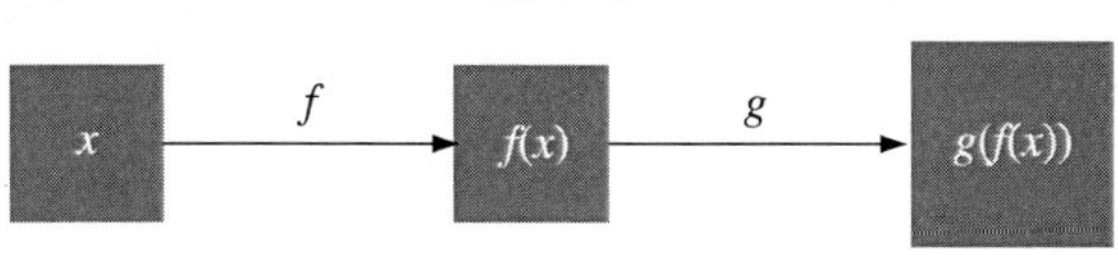

FIGURE 1.12 Composite function.

EXAMPLE 1.4 Let $f : N \rightarrow N, f(x) = 2x + 1$ and $g : N \rightarrow N$, $g(x) = x + 3$. Then

$$fog(x) = f(g(x)) = f(x+3) = 2(x+3) + 3 = 2x + 9 \text{ and}$$

$$gof(x) = g(f(x)) = g(2x+1) = (2x+1) + 3 = 2x + 4$$

1.4 LIMIT OF FUNCTION

While studying the function $f(x)$, the interest is to find the function value near to some point x_0 or value at x_0. Sometimes if there is an irrational number like π or $\sqrt{2}$, then one can only approximate the value of function near to rational numbers. Another scenario happens when a function is not divided by zero and we try to estimate the function value at the point x_0 which will be undefined. For example, $f(x) = \dfrac{x^2-4}{x-2} = x+2, x \neq 2$

This can be understood numerically, *i.e.*,

TABLE 1.1 The value of $f(x)$ w.r.t. x

The value of x	*The value of f(x)*	*The value of x*	*The value of f(x)*
1.9	3.9	2.1	4.1
1.99	3.99	2.01	4.01
1.99999	3.99999	2.000001	4.000001

As x gets closer to 2, $f(x)$ gets closer to 4. Thus, one can understand through definition.

Definition: Let f be a real-valued function defined on an open interval about $x_0 \in R$. If $f(x)$ approaches to the **limit** L as x tend to x_0 and symbolically we write as $x \to x_0$, $f(x) \to L$, *i.e.*, $\lim\limits_{x \to x_0} f(x) = L$.

For the above example, $\lim\limits_{x \to 2} \dfrac{x^2-4}{x-2} = \lim\limits_{x \to 2} x+2 = 2+2 = 4.$

Right-hand and Left-hand Limits

If as $x \to x_0^+$, $f(x) \to l_1$, l_1 is the right-hand limit of a function f at x_0, symbolically we write $\lim\limits_{x \to x_0^+} f(x) = l_1$.

If as $x \to x_0^-$, $f(x) \to l_2$, l_2 is the right-hand limit of a function f at x_0, symbolically we write $\lim\limits_{x \to x_0^-} f(x) = l_2$. According to these two limits, two results were derived:

1. If $\lim\limits_{x \to x_0} f(x) = L$ exists, it must be unique.
2. A function $f(x) \to L$ (tends to L) as $x \to x_0$ if and only if its right-hand limit and left-hand limit exist and those are equal:

$$\lim_{x \to x_0^+} f(x) = L \Leftrightarrow \lim_{x \to x_0^-} f(x) = L \text{ and } \lim_{x \to x_0} f(x) = L$$

EXAMPLE 1.5 $\lim_{x \to 2^+} (4x - 1) = 4(2) - 1 = 7.$

$$\lim_{x \to 2^-} (4x - 1) = 4(2) - 1 = 7 \quad \text{and} \quad \lim_{x \to 2} (4x - 1) = 4(2) - 1 = 7$$

Working Rules of Limit

If for two functions *f*(*x*) and *g*(*x*), $\lim_{x \to x_0} f(x) = L_1$ and $\lim_{x \to x_0} g(x) = L_2$ exists, then

1. $\lim_{x \to x_0} (f(x) + g(x)) = \lim_{x \to x_0} f(x) + \lim_{x \to x_0} g(x) = L_1 + L_2$

2. $\lim_{x \to x_0} (f(x) - g(x)) = \lim_{x \to x_0} f(x) - \lim_{x \to x_0} g(x) = L_1 - L_2$

3. $\lim_{x \to x_0} k.f(x) = k \lim_{x \to x_0} f(x) = kL_1, \; k \in R$

4. $\lim_{x \to x_0} (f(x) \times g(x)) = \lim_{x \to x_0} f(x) \times \lim_{x \to x_0} g(x) = L_1 \times L_2$

5. $\lim_{x \to x_0} \dfrac{f(x)}{g(x)} = \dfrac{\lim_{x \to x_0} f(x)}{\lim_{x \to x_0} g(x)} = \dfrac{L_1}{L_2}; \; g(x) \neq 0, \; \lim_{x \to x_0} g(x) \neq 0$

EXAMPLE 1.6 Let *f*(*x*) = x^2 and *g*(*x*) = $x + 1$,

$$\lim_{x \to 1} (f(x) + g(x)) = \lim_{x \to 1} (x^2 + x + 1) = 1 + 1 + 1 = 3$$

$$\lim_{x \to 1} (f(x) - g(x)) = \lim_{x \to 1} (x^2 - x - 1) = 1 - 1 - 1 = -1$$

$$\lim_{x \to 1} 3 \cdot x^2 = 3 \cdot \lim_{x \to 1} x^2 = 3 \cdot 1 = 3$$

$$\lim_{x \to 1} (f(x) \times g(x)) = \lim_{x \to 1} x^2 \times \lim_{x \to 1} (x + 1) = 1 \times (1 + 1) = 2$$

$$\lim_{x \to 1} \frac{f(x)}{g(x)} = \frac{\lim_{x \to 1} x^2}{\lim_{x \to 1} (x + 1)} = \frac{1}{2}$$

Some Standard Limits

1. The limit of a constant function is constant, *i.e.*, $\lim_{x \to x_0} c = c$, c = constant.

   For example: (i) $\lim_{x \to 1} 0 = 0$; (ii) $\lim_{x \to 3} \sqrt{5} = \sqrt{5}$

2. $\lim\limits_{x \to a} x^n = a^n, n \in R$

For example: (i) $\lim\limits_{x \to 2} x^4 = 2^4 = 16$; (ii) $\lim\limits_{x \to 16} x^{1/2} = \sqrt{16} = 4$

3. $\lim\limits_{x \to a} \dfrac{x^n - a^n}{x - a} = na^{n-1}, n \in R$

For example: (i) $\lim\limits_{x \to 2} \dfrac{x^4 - 16}{x - 2} = \lim\limits_{x \to 2} \dfrac{x^4 - 2^4}{x - 2} = 4 \cdot 2^{4-1} = 4 \cdot 2^3 = 32$

(ii) $\lim\limits_{x \to -1} \dfrac{x^3 + 1}{x + 1} = \lim\limits_{x \to -1} \dfrac{x^3 - (-1)^3}{x - (-1)} = 3 \cdot (-1)^{3-1} = 3$

4. $\lim\limits_{\theta \to 0} \dfrac{\sin \theta}{\theta} = 1$ and $\lim\limits_{\theta \to 0} \dfrac{\tan \theta}{\theta} = 1$

For example: (i) $\lim\limits_{x \to 0} \dfrac{\tan 2x}{x} = \lim\limits_{x \to 0} \dfrac{\tan 2x}{2x} \times 2 = 2 \times \lim\limits_{2x \to 0} \dfrac{\tan 2x}{2x} = 2$

(ii) $$\lim_{x \to 0} \frac{\sin 5x}{\tan 8x} = \frac{\lim\limits_{x \to 0} \dfrac{\sin 5x}{5x} \times 5x}{\lim\limits_{x \to 0} \dfrac{\tan 8x}{8x} \times 8x} = \frac{\lim\limits_{5x \to 0} \dfrac{\sin 5x}{5x}}{\lim\limits_{8x \to 0} \dfrac{\tan 8x}{8x}} \times \frac{\lim\limits_{x \to 0} 5x}{\lim\limits_{x \to 0} 8x} = \frac{1}{1} \times \frac{5}{8} = \frac{5}{8}$$

EXAMPLE 1.7 Evaluate the limits of the following:

1. $$\lim_{x \to 2} \frac{x^2 - 5x + 6}{x^3 - 8} = \lim_{x \to 2} \frac{(x-2)(x-3)}{(x-2)(x^2 + 2x + 4)}$$
$$= \lim_{x \to 2} \frac{x - 3}{x^2 + 2x + 4} = \frac{2 - 3}{2^2 + 2(2) + 4} = \frac{-1}{12}$$

2. $$\lim_{x \to 2} \frac{x^4 - 16}{x^3 - 8} = \frac{\lim\limits_{x \to 2} \dfrac{x^4 - (2)^4}{x - 2}}{\lim\limits_{x \to 2} \dfrac{x^3 - (2)^3}{x - 2}} = \frac{4 \cdot (2)^{4-1}}{3 \cdot (2)^{3-1}} = \frac{32}{12} = \frac{8}{3}$$

3. $$\lim_{x \to 3} \frac{\sqrt{x+2} - \sqrt{5}}{x - 3} = \lim_{x \to 3} \frac{\sqrt{x+2} - \sqrt{5}}{x - 3} \times \frac{\sqrt{x+2} + \sqrt{5}}{\sqrt{x+2} + \sqrt{5}}$$
$$= \lim_{x \to 3} \frac{x + 2 - 5}{x - 3 \times \left(\sqrt{x+2} + \sqrt{5}\right)}$$
$$= \lim_{x \to 3} \frac{x - 3}{x - 3 \times \left(\sqrt{x+2} + \sqrt{5}\right)}$$
$$= \lim_{x \to 3} \frac{1}{\sqrt{x+2} + \sqrt{5}} = \frac{1}{2\sqrt{5}}$$

4. $\lim_{x\to 0}\dfrac{\cos x-1}{x}=\lim_{x\to 0}\dfrac{-2\sin^2(x/2)}{x}=-\lim_{x/2\to 0}\dfrac{\sin(x/2)}{x/2}\times\sin(x/2)$

$$=-\lim_{x/2\to 0}\frac{\sin(x/2)}{x/2}\times\lim_{x/2\to 0}\sin(x/2)=-1\times 0=0$$

1.5 INFINITE LIMITS AND LIMITS AT INFINITY

Let f be a real-valued function with a limit greater than some number (say k) however large, then one can say that x tends to infinity *i.e.*, $\lim_{x\to\infty}\left(\dfrac{1}{x}\right)=0$.

1. $\lim_{x\to\infty}\left(\dfrac{x+1}{x^3}\right)=\lim_{x\to\infty}\left(\dfrac{1}{x^2}+\dfrac{1}{x^3}\right)=\lim_{x\to\infty}\dfrac{1}{x^2}+\lim_{x\to\infty}\dfrac{1}{x^3}=0+0=0$.

2. $\lim_{x\to\infty}\left(\dfrac{x^3+3x+4}{4x^3+x^2+1}\right)=\lim_{x\to\infty}\left(\dfrac{\dfrac{x^3+3x+4}{x^3}}{\dfrac{4x^3+x^2+1}{x^3}}\right)$

$$=\lim_{x\to\infty}\left(\frac{1+\dfrac{3}{x^2}+\dfrac{4}{x^3}}{4+\dfrac{1}{x}+\dfrac{1}{x^3}}\right)=\frac{1+0+0}{4+0+0}=\frac{1}{4}$$

Some Standard Results

1. $\lim_{x\to 0}(1+x)^{\frac{1}{x}}=e$

2. $\lim_{x\to\infty}\left(1+\dfrac{1}{x}\right)^x=e$

3. $\lim_{x\to 0}\dfrac{a^x-1}{x}=\log_e a,\ a\in R^+-\{1\}$

4. $\lim_{x\to 0}\left(\dfrac{e^x-1}{x}\right)=1$

EXAMPLE 1.8 Evaluate the limits of the following:

1. $$\lim_{n\to\infty}\left(\frac{\Sigma n}{(n+1)(3n+2)}\right)=\lim_{n\to\infty}\left(\frac{\frac{n(n+1)}{2}}{(n+1)\,(3n+2)}\right)$$

$$=\lim_{n\to\infty}\left(\frac{n}{2(3n+2)}\right)=\lim_{n\to\infty}\left(\frac{1}{2\left(3+\frac{2}{n}\right)}\right)=\frac{1}{6+0}=\frac{1}{6}$$

2. $$\lim_{x\to 0}(1+2x)^{1+\frac{1}{x}}=\lim_{x\to 0}(1+2x)^1\times\lim_{x\to 0}(1+2x)^{\frac{1}{x}}$$

$$(1+2(0))\times\lim_{x\to 0}(1+2x)^{\frac{1}{2x}\times 2}=(1+0)\times\left\{\lim_{x\to 0}(1+2x)^{\frac{1}{2x}}\right\}^2=1\times e^2=e^2$$

3. $$\lim_{x\to 0}\frac{8^x-5^x}{x}=\lim_{x\to 0}\frac{(8^x-1)-(5^x-1)}{x}=\lim_{x\to 0}\frac{(8^x-1)}{x}-\lim_{x\to 0}\frac{(5^x-1)}{x}$$

$$=\log 8-\log 5=\log\left(\frac{8}{5}\right)$$

1.6 CONTINUOUS FUNCTIONS

We know that graph of any function can be sketched without lifting a pen in its domain is considered as a continuous function. Outputs of this function moving continuously according to its inputs, cover all the values in between without jumping from one point to another. We can check continuity at three different levels:

1. *Continuity at a Point*

A function $y = f(x)$ is said to be continuous at $x = x_0$ of its domain, if $\lim_{x\to x_0} f(x)=f(x_0)$. That is, $\lim_{x\to x_0^+} f(x)=\lim_{x\to x_0^-} f(x)=\lim_{x\to x_0} f(x)=f(x_0)$.

EXAMPLE 1.9 $f(x)=\begin{cases}\dfrac{2x^2-3x-2}{x-2} & x\neq 2\\ 5 & x=2\end{cases}$ then check the continuity of $x = 2$.

Here,

$$\lim_{x \to 2^-} \left(\frac{2x^2 - 3x - 2}{x - 2} \right) = \lim_{x \to 0} \left(\frac{2(2-h)^2 - 3(2-h) - 2}{(2-h) - 2} \right)$$

$$= \lim_{h \to 0} \left(\frac{2h^2 - 5h}{-h} \right) = \lim_{h \to 0} \frac{h(2h - 5)}{-h} = 5$$

$$\lim_{x \to 2^+} \left(\frac{2x^2 - 3x - 2}{x - 2} \right) = \lim_{h \to 0} \left(\frac{2(2+h)^2 - 3(2+h) - 2}{(2+h) - 2} \right)$$

$$= \lim_{h \to 0} \left(\frac{2h^2 + 5h}{h} \right) = \lim_{h \to 0} \frac{h(2h + 5)}{h} = 5 \text{ and } f(2) = 5.$$

Thus, $\lim\limits_{x \to 2^+} f(2) = \lim\limits_{x \to 2^-} f(2) = f(2)$.

Hence, $f(x)$ is continuous at $x = 2$.

2. *Continuity in an Open Interval*

A function $y = f(x)$ is said to be continuous in an open interval (a, b) if it is continuous at each point of (a, b).

EXAMPLE 1.10 The function $y = \lfloor x \rfloor$ is continuous in (1, 2).

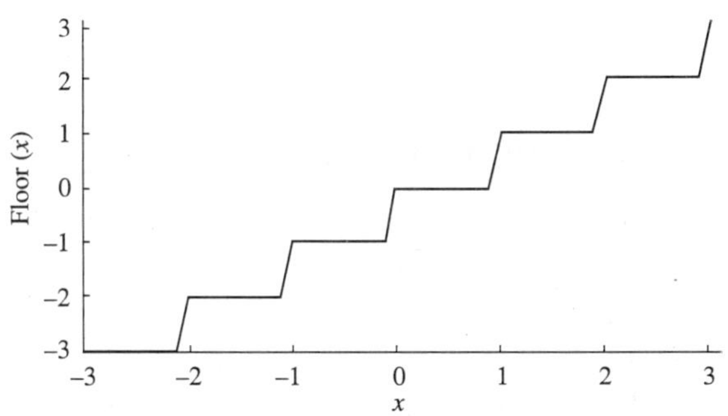

FIGURE 1.13 Graph of $y = \lfloor x \rfloor$.

From the graph, we can see that the function is continuous at every point except the integer value *i.e.*, it is continuous in (a, b), $a, b \in Z$.

3. *Continuity in a Closed Interval*

A function $y = f(x)$ is said to be continuous in an open interval $[a, b]$ if,

(i) it is continuous in (a, b).

(ii) The right-hand limit at a point a is equal to the value of the function at a, $\lim\limits_{x \to a^+} f(x) = f(a)$.

(iii) The left-hand limit at a point b is equal to the value of the function at b, $\lim\limits_{x \to b^-} f(x) = f(b)$.

EXAMPLE 1.11 The function $y=\sqrt{9-x^2}$ is continuous at every point in [–3, 3]. It is also right continuous at $x = -2$ and left continuous at $x = 2$.

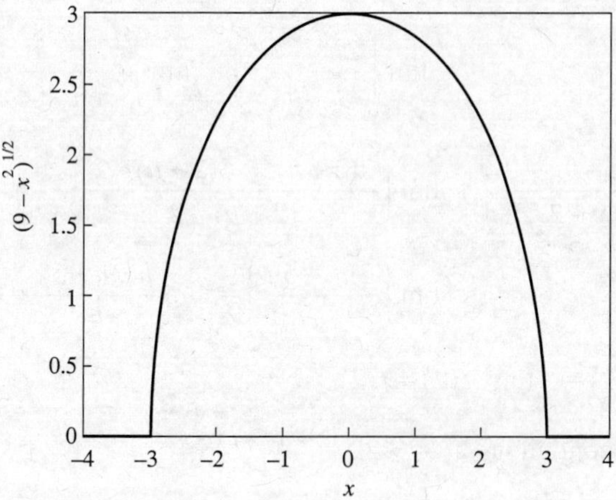

FIGURE 1.14 Graph of $y=\sqrt{9-x^2}$.

The polynomial function and rational function (denominator not equal to zero) are continuous at every point on the real line.

Functions like modulus, logarithmic, trigonometric, and exponentials are continuous in their specified domain.

1.7 DISCONTINUOUS FUNCTIONS

There are types of discontinuities:

A function $f(x)$ has a **removable discontinuity** at a point x_0 if the left-hand limit and right-hand limit at a point are not equal to $f(x_0)$.

If the left-hand limit and right-hand limit at a point exist but they are not equal then the function has **discontinuity of the first kind** or **jump discontinuity**, none of the limits exist then $f(x)$ is said to have **discontinuity of the second kind** and either or both of the limits be infinite then $f(x)$ has an **infinite discontinuity**.

EXAMPLE 1.12 $f(x)$ = [sin x]. Check the discontinuity at $x=\dfrac{\pi}{2}$.

Here, $\lim\limits_{x\to\pi/2}[\sin x]=0$ but $f(\pi/2) = 1$. To remove this discontinuity, we can redefine $f(x)$ as, $f(x)=\begin{cases}[\sin x], & x\in\left(0,\dfrac{\pi}{2}\left(\dfrac{\pi}{2},\pi\right)\right)\\ 0 & x=\dfrac{\pi}{2}\end{cases}$. Now, $f(x)$ is continuous for all $x \in (0, \pi)$. Thus, the function has a removable discontinuity.

EXAMPLE 1.13 $f(x) = [x]$. Check the discontinuity at $x = 4$.

$$\text{LHL} = \lim_{x\to 4^-} [x] = 3 \text{ and RHL} = \lim_{x\to 4^+} [x] = 4$$

LHL ≠ RHL. Hence, the function is discontinuous of the first kind at $x = 4$.

EXAMPLE 1.14 $f(x) = e^{1/x}$. Check the discontinuity at $x = 0$.

In this example, $\lim_{x\to 0^-} e^{1/x} = 0$ and $\lim_{x\to 0^+} e^{1/x} = \infty$. Hence, the function has an infinite discontinuity at $x = 0$.

EXAMPLE 1.15 $f(x) = \dfrac{x}{1-x}$. Check the discontinuity at $x = 1$.

Here, neither of the limits exists. Hence, the function is discontinuous of the second kind at $x = 1$.

1.8 OPERATIONS ON CONTINUOUS FUNCTIONS

If the functions $f(x)$ and $g(x)$ are continuous at a point $x = x_0$, then the following combinations are also continuous at $x = x_0$ and operations hold:

1. Addition: $f(x) + g(x)$
2. Differences: $f(x) - g(x)$
3. Products: $f(x) \cdot g(x)$
4. Quotients: $f(x)/g(x)$
5. Constant Multiple: $k \cdot f(x)$, k is any number
6. Powers: $(f(x))^n$, n is a positive number
7. Roots: $\sqrt[n]{f(x)}$, n is a positive integer

In the composition of two functions $g(f(x)) = g \circ f$ is continuous if f is continuous at $x = x_0$ and g is continuous at $f(x_0)$.

EXERCISES

Find the domain and range of the following functions: (1–3):

1. $f(x) = 1 + x^2$
2. $f(x) = \dfrac{4}{3-x}$
3. $f(x) = \sqrt{4-x}$
4. If $f: R \to R$ defined by $f(x) = 2x + 1$ and $g: R \to R$ defined by $g(x) = 3x - 2$. Find *fog*, *gof*, *fof* and *gog*.

5. Complete the blanks:

$f(x)$	$g(x)$	$f(g(x))$
$\sqrt{x}$	$x-7$	?
$\dfrac{x}{x-1}$	$\dfrac{x}{x-1}$	?
$\sqrt{x-5}$	?	$\sqrt{x^2-5}$

Evaluate the following limits: (6–17):

6. $\lim\limits_{x\to -2}(2x^3-3x^2+4x+5)$

7. $\lim\limits_{x\to 0}\left(\dfrac{3}{\sqrt{4x+1}+2}\right)$

8. $\lim\limits_{x\to 3}\left(\dfrac{x-3}{x^2-9}\right)$

9. $\lim\limits_{x\to -4}\left(\dfrac{x^2-3x-28}{x^2+7x+12}\right)$

10. $\lim\limits_{x\to 2}\left(\dfrac{\sqrt{x+7}-3}{x-2}\right)$

11. $\lim\limits_{x\to 4}\left(\dfrac{4-x}{5-\sqrt{x^2+9}}\right)$

12. $\lim\limits_{x\to 1}\left(\dfrac{\sqrt{9x+7}-\sqrt{3x+13}}{\sqrt{3x+1}-\sqrt{5x-1}}\right)$

13. $\lim\limits_{x\to 0}(2\sin x-\cos x)$

14. $\lim\limits_{x\to 0}(x^2+3)(4+\sec x)$

15. $\lim\limits_{x\to 0}\left(\dfrac{\sin 2x(1-\cos 2x)}{x^3}\right)$

16. $\lim\limits_{x\to 0}\left(\dfrac{\sin 2x+\tan 3x}{4x-\tan 5x}\right)$

17. $\lim\limits_{x\to 0}\left(\dfrac{\sin(2+x)-\sin(2-x)}{x}\right)$

Evaluate the following limits: (18–24):

18. $\lim\limits_{x\to \infty}\left(\dfrac{6x^3-4x^2+3x+5}{3x^3+2x^2-7x+1}\right)$

19. $\lim\limits_{n\to \infty}\left(\dfrac{\sum n^2}{\sum n^3}\right)$

20. $\lim\limits_{x\to \infty}\left(\dfrac{(3x+2)^{40}\cdot(4+x)^{10}}{(3x+7)^{50}}\right)$

21. $\lim\limits_{x\to \infty}\left(\dfrac{x+4}{x-4}\right)^x$

22. $\lim\limits_{x\to 0}\left(\dfrac{2x+1}{1-5x}\right)^{\frac{1}{x}}$

23. $\lim\limits_{x\to 0}\left(\dfrac{15^x-5^x-3^x+1}{x^2}\right)$

24. $\lim\limits_{x\to 0}\left(\dfrac{e^{3x}-e^{-3x}}{x}\right)$

25. Discuss the continuity of the function: $f(x)=\begin{cases}1 & x>0\\ 0 & x=0\\ -1 & x<0\end{cases}$.

26. Examine continuity of function: $f(x)=\begin{cases}\dfrac{x^2-x-6}{x-3}, & x\neq 3\\ 5, & x=3\end{cases}$.

27. If $\lim\limits_{x\to 4}\left(\dfrac{f(x)-6}{x-4}\right)=1$, find $\lim\limits_{x\to 4} f(x)$.

28. Discuss the continuity of the function: $f(x)=\begin{cases}\dfrac{x^2}{3}-3, & x\leq 3\\ 3-\dfrac{9}{x}, & x>3\end{cases}$.

29. For what value of a is, $f(x)=\begin{cases}a^2x-2a, & x\geq 2\\ 12, & x<2\end{cases}$.

30. For what value of a and b is, $f(x)=\begin{cases}-2, & x\leq -1\\ ax-b & -1\leq x<1\\ 3, & x\geq 1\end{cases}$.

Answers

1. $D:(-\infty,\infty)$ $R:[1,\infty)$
2. $D:(-\infty,3)\cup(3,\infty)$ $R:(-\infty,0)\cup(0,\infty)$
3. $D:(-\infty,4]$ $R:[0,\infty)$
4. $fog = 6x-3$, $gof = 6x+1$, $fof = 4x+3$, $gog = 9x-8$
5. $\sqrt{x-7}, x, x^2$
6. –31
7. 1
8. 6
9. 11
10. 1/6
11. 5/4
12. –3/2
13. –1
14. 15
15. 4
16. –5
17. 2 cos 2
18. 2
19. 0
20. $1/3^{10}$
21. e^8
22. e^7
23. (log 5) (log 3)
24. 6
25. Discontinuous
26. Continuous for all x
27. 6
28. Continuous
29. (–2, 3)
30. $a = 5/2$, $b = -1/2$

2 Differentiation

2.1 INTRODUCTION

If we are starting the history of calculus, two legendary people Sir Issac Newton (1642–1727) and G.W. Leibnitz (1646–1717) always come to mind. Their independent research gave a new direction to Mathematics during the 17th century.

The derivative of a function is the instantaneous rate of change of a function. We define the slope and tangent to a curve at a point and the derivative to that point.

To find the slope or tangent of the curve $y = f(x)$ through $P(x_0, f(x_0))$ and $Q(x_0 + h, fE(x_0 + h))$. Then we calculate the slope as $h \to 0$. Hence the slope m will be, $m = \lim_{h \to 0} \frac{f(x_0 + h) - f(x_0)}{h}$, $h \neq 0$.

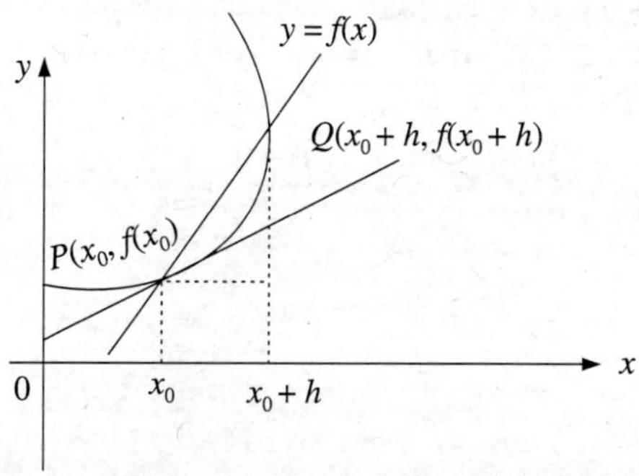

FIGURE 2.1 Tangent of the curve.

If this limit $m = \lim_{h \to 0} \frac{f(x_0 + h) - f(x_0)}{h}$ exists, then it is said to be **derivative or differential co-efficient** of $f(x)$ with respect to x. It is denoted by $f'(x)$ or $\frac{d}{dx}(f(x))$ or $\frac{dy}{dx}$ or y_1.

EXAMPLE 2.1 Using the definition, the derivative of $f(x) = x$ with respect to x.

Here, $f(x) = x$ and $f(x + h) = x + h$.

By definition, $f'(x) = \lim_{h \to 0} \frac{f(x+h) - f(x)}{h}$

$$\Rightarrow f'(x) = \lim_{h \to 0} \frac{(x+h)-(x)}{h} = \lim_{h \to 0} \frac{x+h-x}{h} = \lim_{h \to 0} \frac{h}{h} = \lim_{h \to 0} (1) = 1$$

EXAMPLE 2.2 Using the definition, the derivative of $f(x) = \sqrt{x}$ with respect to x.

Here, $f(x) = \sqrt{x}$ and $f(x+h) = \sqrt{x+h}$.

By definition, $f'(x) = \lim_{h \to 0} \frac{f(x+h) - f(x)}{h}$

$$\therefore\ f'(x) = \lim_{h \to 0} \frac{\left(\sqrt{x+h}\right) - \left(\sqrt{x}\right)}{h} = \lim_{h \to 0} \frac{\left(\sqrt{x+h}\right) - \left(\sqrt{x}\right)}{h} \times \frac{\left(\sqrt{x+h}\right) + \left(\sqrt{x}\right)}{\left(\sqrt{x+h}\right) + \left(\sqrt{x}\right)}$$

$$\therefore\ f'(x) = \lim_{h \to 0} \frac{(x+h)-(x)}{h\left(\sqrt{x+h} + \sqrt{x}\right)} = \lim_{h \to 0} \frac{1}{\sqrt{x+h} + \sqrt{x}} = \frac{1}{\sqrt{x+0} + \sqrt{x}} = \frac{1}{2\sqrt{x}}$$

EXAMPLE 2.3 Differentiate $f(x) = \frac{x}{x-1}$ using the definition.

In this example, $f(x) = \frac{x}{x-1}$ and $f(x+h) = \frac{x+h}{x+h-1}$.

From definition, $f'(x) = \lim_{h \to 0} \frac{f(x+h) - f(x)}{h} = \lim_{h \to 0} \frac{\frac{x+h}{x+h-1} - \frac{x}{x-1}}{h}$

$$\therefore\ f'(x) = \lim_{h \to 0} \frac{x+h(x-1) - x(x+h-1)}{h(x+h-1)\,(x-1)} = \lim_{h \to 0} \frac{x^2 - x + hx - h - x^2 - xh + x}{h(x+h-1)\,(x-1)}$$

$$\therefore\ f'(x) = \lim_{h \to 0} \frac{-h}{(h)(x+h-1)(x-1)} = \frac{-1}{(x-1)\,(x-1)} = \frac{-1}{(x-1)^2}$$

2.2 DERIVATIVES OF SOME WELL-DEFINED FUNCTION

1. The derivative of any constant function k is zero, $\frac{d}{dx}(k) = 0$.
2. $\frac{d}{dx}(x^n) = nx^{n-1},\ n \in R$.

For example, (i) $\dfrac{d}{dx}\left(x^{\frac{4}{3}}\right) = \dfrac{4}{3}x^{\frac{4}{3}-1} = \dfrac{4}{3}x^{\frac{1}{3}}$

(ii) $\dfrac{d}{dx}(x) = (1)x^{1-1} = 1$

(iii) $\dfrac{d}{dx}(\sqrt{x}) = \dfrac{d}{dx}\left(x^{\frac{1}{2}}\right) = \dfrac{1}{2}x^{-\frac{1}{2}} = \dfrac{1}{2\sqrt{x}}$

(iv) $\dfrac{d}{dx}\left(\dfrac{1}{x}\right) = \dfrac{d}{dx}(x^{-1}) = (-1)x^{-2} = \dfrac{-1}{x^2}$

3. $\dfrac{d}{dx}(|x|) = \dfrac{x}{|x|},\ x \neq 0$

4. $\dfrac{d}{dx}(a^x) = a^x \cdot \log_e a,\ a \in R^+;\ x \in R$

For example, $\dfrac{d}{dx}(3^x) = 3^x \cdot \log_e 3$

5. $\dfrac{d}{dx}(e^x) = e^x, \forall x \in R$

6. $\dfrac{d}{dx}(\log_e x) = \dfrac{1}{x},\ \forall x \in R^+$

7. Derivative of Trigonometric functions:

(i) $\dfrac{d}{dx}(\sin x) = \cos x$ (ii) $\dfrac{d}{dx}(\sin^{-1} x) = \dfrac{1}{\sqrt{1-x^2}}$

(iii) $\dfrac{d}{dx}(\cos x) = -\sin x$ (iv) $\dfrac{d}{dx}(\cos^{-1} x) = -\dfrac{1}{\sqrt{1-x^2}}$

(v) $\dfrac{d}{dx}(\tan x) = \sec^2 x$ (vi) $\dfrac{d}{dx}(\tan^{-1} x) = \dfrac{1}{1+x^2}$

(vii) $\dfrac{d}{dx}(\cot x) = -\operatorname{cosec}^2 x$ (viii) $\dfrac{d}{dx}(\cot^{-1} x) = -\dfrac{1}{1+x^2}$

(ix) $\dfrac{d}{dx}(\sec x) = \sec x \tan x$ (x) $\dfrac{d}{dx}(\sec^{-1} x) = \dfrac{1}{|x|\sqrt{x^2-1}}$

(xi) $\dfrac{d}{dx}(\operatorname{cosec} x) = -\operatorname{cosec} x \cot x$ (xii) $\dfrac{d}{dx}(\operatorname{cosec}^{-1} x) = -\dfrac{1}{|x|\sqrt{x^2-1}}$

The Working Rules for Differentiation:

If $f(x)$ and $g(x)$ are differentiable at x then,

1. Sum and difference Rule: $f \pm g$ is differentiable, *i.e.*,

$$\frac{d}{dx}(f(x) \pm g(x)) = \frac{d}{dx}(f(x)) \pm \frac{d}{dx}(g(x)).$$

For example, if $y = x^3 + 3^x - \cos x$ then find $\frac{dy}{dx}$.

$$\frac{dy}{dx} = \frac{d(x^3 + 3^x - \cos x)}{dx} = \frac{d}{dx}(x^3) + \frac{d}{dx}(3^x) - \frac{d}{dx}(\cos x)$$

$$\therefore \quad \frac{dy}{dx} = 3x^2 + 3^x \log_e 3 + \sin x$$

2. $\frac{d}{dx}(k \cdot f(x)) = k \cdot \frac{d}{dx}(f(x)), k \in R$

3. Product Rule: $f \cdot g$ is differentiable, *i.e.*,

$$\frac{d}{dx}(f(x)\, g(x)) = f(x) \cdot \frac{d}{dx}(g(x)) + g(x) \cdot \frac{d}{dx}(f(x))$$

For example, if $y = x^3 \cdot \cos x$ then find $\frac{dy}{dx}$.

$$\frac{dy}{dx} = \frac{d}{dx}(x^3 \cdot \cos x) = x^3 \frac{d}{dx}(\cos x) + \cos x \frac{d}{dx}(x^3)$$

$$\therefore \quad \frac{dy}{dx} = x^3(-\sin x) + \cos x(3x^2) = x^2(-\sin x \cdot x + 3\cos x)$$

4. Division Rule: f/g is differentiable, *i.e.*,

$$\frac{d}{dx}\left(\frac{f(x)}{g(x)}\right) = \frac{g(x) \cdot \frac{d}{dx}(f(x)) - f(x) \cdot \frac{d}{dx}(g(x))}{[g(x)]^2}, \; g(x) \neq 0$$

For example, if $y = \frac{x^3}{e^x}$ then find $\frac{dy}{dx}$.

$$\frac{dy}{dx} = \frac{d}{dx}\left(\frac{x^3}{e^x}\right) = \frac{e^x \frac{d}{dx}(x^3) - x^3 \frac{d}{dx}(e^x)}{[e^x]^2}$$

$$\therefore \quad \frac{dy}{dx} = \frac{e^x(3x^2) - x^3 e^x}{e^{2x}} = \frac{e^x x^2(3-x)}{e^{2x}} = \frac{x^2(3-x)}{e^x}$$

EXAMPLE 2.4 Differentiate the following functions:

1. $y = \dfrac{7x^4 + 2x^3 - x^2 + 2}{4x^2}$

Here, $\dfrac{dy}{dx} = \dfrac{d}{dx}\left(\dfrac{7x^4 + 2x^3 - x^2 + 2}{4x^2}\right) = \dfrac{d}{dx}\left(\dfrac{7x^4}{4x^2} + \dfrac{2x^3}{4x^2} - \dfrac{x^2}{4x^2} + \dfrac{2}{4x^2}\right)$

$$\therefore \frac{dy}{dx} = \frac{d}{dx}\left(\frac{7x^2}{4} + \frac{x}{2} - \frac{1}{4} + \frac{1}{2x^2}\right) = \frac{7}{4}\frac{d(x^2)}{dx} + \frac{1}{2}\frac{d(x)}{dx} - \frac{d}{dx}\left(\frac{1}{4}\right) + \frac{1}{2}\frac{d(x^{-2})}{dx}$$

$$\therefore \frac{dy}{dx} = \frac{7}{4}(2x) + \frac{1}{2}(1) - 0 + \frac{1}{2}(-2x^{-3}) = \frac{7x}{2} + \frac{1}{2} - \frac{1}{x^3}$$

2. $y = \log x \tan x + x^3$

Here, $\dfrac{dy}{dx} = \dfrac{d}{dx}(\log x \sin x + x^3) = \log x \dfrac{d}{dx}(\tan x) + \tan x \dfrac{d}{dx}(\log x) + \dfrac{d}{dx}(x^3)$

$$\therefore \frac{dy}{dx} = \log x \cdot \sec^2 x + \tan x \cdot \frac{1}{x} + 3x^2$$

3. $y = \dfrac{t^2 - 1}{t^3 + 1}$

$$\frac{dy}{dt} = \frac{d}{dt}\left(\frac{t^2 - 1}{t^3 + 1}\right) = \frac{(t^3 + 1)\dfrac{d}{dt}(t^2 - 1) - (t^2 - 1)\dfrac{d}{dt}(t^3 + 1)}{[t^3 + 1]^2}$$

$$\therefore \frac{dy}{dt} = \frac{(t^3 + 1)(2t - 0) - (t^2 - 1)(3t^2 + 0)}{[t^3 + 1]^2}$$

$$\therefore \frac{dy}{dt} = \frac{(2t^4 + 2t) - (3t^4 - 3t^2)}{(t^3 + 1)^2} = \frac{-t^4 + 3t^2 + 2t}{(t^3 + 1)^2}$$

THEOREM 2.1 Differentiable functions are continuous: If f has a derivative at $x = c$, then f is continuous at $x = c$.

Proof: We have given that, f has a derivative at $x = c$, *i.e.*, $f'(c)$ exists, we must show that $\lim_{x \to c} f(x) = f(c)$ or $\lim_{h \to 0} f(c + h) = f(c)$. If $h \neq 0$, then

$$f(c + h) = f(c) + (f(c + h) - f(c))$$

$$= f(c) + \left(\frac{f(c + h) - f(c)}{h} \cdot h\right)$$

Now, take the limit as $h \to 0$. By the rule of limit, we have,

$$\lim_{h \to 0} f(c + h) = \lim_{h \to 0} f(c) + \lim_{h \to 0}\left(\frac{f(c + h) - f(c)}{h}\right) \cdot \lim_{h \to 0} h$$

$$= \lim_{h \to 0} f(c) + f'(c) \cdot 0$$

$$= f(c).$$

Thus, f is continuous from the right side of $x = c$. With similar arguments, f is continuous from the left side of $x = c$. Hence, f is continuous at $x = c$.

Chain Rule: If $f(y)$ is a differentiable function of $y = g(x)$ and $y = g(x)$ is the differentiable function at x, then the derivative of f with respect to x is defined as, $\frac{df}{dx} = \frac{df}{dy} \times \frac{dy}{dx}$. It is known as the chain rule for differentiation or can say derivative of a composite function.

EXAMPLE 2.5 If $y = (\log_e x)^3$ then find $\frac{dy}{dx}$.

Here, $y = (\log_e x)^3$. Let $p = \log_e x \Rightarrow \frac{dp}{dx} = \frac{1}{x}$

and $y = p^3 \Rightarrow \frac{dy}{dp} = 3p^2$

From the chain rule, $\frac{dy}{dx} = \frac{dy}{dp} \times \frac{dp}{dx} = 3p^2 \times \frac{1}{x} = \frac{3(\log_e x)^2}{x}$

EXAMPLE 2.6 If $y = \sin(2^x x^2)$ then find $\frac{dy}{dx}$.

Here, $y = \sin(2^x x^2)$. Let $p = 2^x x^2 \Rightarrow \frac{dp}{dx} = 2^x(2x) + x^2(2^x \log_e 2)$

and $y = \sin p \Rightarrow \frac{dy}{dp} = \cos p$

From the chain rule, $\frac{dy}{dx} = \frac{dy}{dp} \times \frac{dp}{dx} = \cos p \times (2^x(2x) + x^2(2^x \log_e 2))$

$$\frac{dy}{dx} = \cos(2^x x^2) \times (2^x(2x) + x^2(2^x \log_e 2))$$

One can use the short-cut method, $y = \sin(2^x x^2)$.

Now firstly take the derivative of the outer function, *i.e.*, $\cos(...) = \cos(2^x x^2)$ and then take the derivative of the inner function *i.e.*, $2^x(2x) + x^2(2^x \log_e 2)$.

$$\therefore \qquad \frac{dy}{dx} = \cos(2^x x^2) \times (2^x(2x) + x^2(2^x \log_e 2)).$$

EXAMPLE 2.7 If $y = \log(\sin x^2)$ then find $\frac{dy}{dx}$.

$$\frac{dy}{dx} = \frac{d}{dx}(\log(\sin x^2)) = \frac{1}{\sin x^2} \times \frac{d}{dx}(\sin x^2) = \frac{1}{\sin x^2} \times \cos x^2 \times \frac{d}{dx}(x^2)$$

$$\therefore \quad \frac{dy}{dx} = \frac{1}{\sin x^2} \times \cos x^2 \times 2x = 2x \cot x^2 .$$

EXAMPLE 2.8 If $y = (x^3 + \sec x)^3$ then find $\frac{dy}{dx}$.

$$\frac{dy}{dx} = \frac{d}{dx}((x^3 + \sec x)^3) = 3(x^3 + \sec x)^2 \times \frac{d}{dx}(x^3 + \sec x)$$

$$= 3(x^3 + \sec x)^2 (3x^2 + \sec x \tan x)$$

2.3 SUCCESSIVE DIFFERENTIATION

The function $y = f(x)$ is defined. The derivative of a function $f'(x) = \frac{dy}{dx}$ is also a function, then it will also possess a derivative, we denote it by $f''(x) = \frac{d}{dx}\left(\frac{dy}{dx}\right) = \frac{d^2y}{dx^2}$ known as the second derivative of a function $f(x)$.

We can further repeat the process for derivatives $f'''(x) = \frac{d}{dx}\left(\frac{d^2y}{dx^2}\right) = \frac{d^3y}{dx^3}$ and so on. Thus, the successive derivatives of $y = f(x)$ can be denoted by symbols $f'(x), f''(x), f'''(x), \ldots$ or $y_1, y_2, y_3, \ldots$ or $\frac{dy}{dx}, \frac{d^2y}{dx^2}, \frac{d^3y}{dx^3}, \ldots$. Thus, it is the process of differentiation of a function successively to derive its higher order derivatives.

EXAMPLE 2.9 If $y = e^x \cos x$ find $\frac{d^2y}{dx^2}$.

We have, $y = e^x \cos x \Rightarrow \frac{dy}{dx} = e^x(-\sin x) + \cos x\, e^x = e^x(\cos x - \sin x)$.

$$\Rightarrow \frac{d^2y}{dx^2} = \frac{d}{dx}\left(\frac{dy}{dx}\right) = \frac{d}{dx}(e^x(\cos x - \sin x))$$

$$\Rightarrow \frac{d^2y}{dx^2} = e^x(-\sin x - \cos x) + (\cos x - \sin x)e^x$$

$$\Rightarrow \frac{d^2y}{dx^2} = e^x(-\sin x - \cos x + \cos x - \sin x) = -2e^x \sin x$$

EXAMPLE 2.10 If $y = x - \cot x$ then prove that $\sin^2 x \frac{d^2y}{dx^2} - 2y + 2x = 0$.

Here, $y = x - \cot x$ differentiating w.r.t. x we have,

$$\frac{dy}{dx} = 1 + \text{cosec}^2 x$$

Again differentiating w.r.t. x we have,

$$\frac{d^2y}{dx^2} = 2\ \text{cosec}\ x \cdot (-\ \text{cosec}\ x \cdot \cot x) = -2\,\text{cosec}^2 x \cdot \cot x$$

$$\text{LHS} = \sin^2 x \frac{d^2y}{dx^2} - 2y + 2x$$

$$= \sin^2 x(-2\ \text{cosec}^2 x \cdot \cot x) - 2(x - \cot x) + 2x$$

$$= -2\,\text{cosec}^2 x \cdot \sin^2 x \cdot \cot x - 2x + 2\cot x + 2x = 0 = \text{RHS}$$

EXAMPLE 2.11 If $y = a\cos(\log x) + b\sin(\log x)$; a, b are constants, prove that $x^2 y_2 + xy_1 + y = 0$.

Here, $y = a\cos(\log x) + b\sin(\log x)$ differentiating w.r.t. x we have,

$$y_1 = -a\sin(\log x)\cdot\frac{1}{x} + b\cos(\log x)\cdot\frac{1}{x} = \frac{1}{x}(-a\sin(\log x) + b\cos(\log x))$$

and,

$$y_2 = \frac{-1}{x^2}(-a\sin(\log x) + b\cos(\log x)) + \frac{1}{x}\left(-a\cos(\log x)\cdot\frac{1}{x} - b\sin(\log x)\cdot\frac{1}{x}\right)$$

$$y_2 = \frac{-1}{x^2}((-a+b)\sin(\log x) + (b+a)\cos(\log x))$$

$$= \frac{1}{x^2}((a-b)\sin(\log x) - (a+b)\cos(\log x))$$

Now, LHS $= x^2 y_2 + xy_1 + y$

$$= x^2\left[\frac{1}{x^2}((a-b)\sin(\log x) - (a+b)\cos(\log x))\right] + x\left[\frac{1}{x}(-a\sin(\log x) + b\cos(\log x))\right]$$

$$+ a\cos(\log x) + b\sin(\log x)$$

$= 0 =$ RHS

2.4 n^{th}-ORDER DERIVATIVES OF STANDARD FUNCTION

1. **n^{th} order derivatives of x^m, where $m \geq n$ is a positive integer**

Here, $y = x^m$ differentiating w.r.t. x,

$y_1 = mx^{m-1}$

$y_2 = m(m-1)x^{m-2}$

$y_3 = m(m-1)(m-2)x^{m-3}$

... ...

Hence, in general $y_n = m(m-1)(m-2)(m-3)\ldots(m-(n-1))x^{m-n}$

If we divide and multiply with the term $(m-n)(m-(n+1))\ldots 3.2.1$,

$$y_n = \frac{m(m-1)(m-2)(m-3)\ldots(m-(n-1))(m-n)(m-(n+1))\ldots 3.2.1}{(m-n)(m-(n+1))\ldots 3.2.1}x^{m-n}$$

$$\therefore \quad \boxed{y_n = \frac{m!}{(m-n)!}x^{m-n}}$$

One can say that, if $y = x^n$, then $y_n = n!$.

2. **n^{th} order derivatives of $(ax+b)^m$, where $m \geq n$ is a positive integer**

Let, $y = (ax+b)^m$ differentiating w.r.t. x,

$y_1 = ma(ax+b)^{m-1}$

$y_2 = m(m-1)a^2(ax+b)^{m-2}$

$y_3 = m(m-1)(m-2)a^3(ax+b)^{m-3}$

... ...

$y_n = m(m-1)(m-2)(m-3)\ldots(m-(n-1))a^n(ax+b)^{m-n}$

Similar way,

$$y_n = \frac{m(m-1)(m-2)(m-3)\ldots(m-(n-1))(m-n)(m-(n+1))\ldots 3.2.1}{(m-n)(m-(n+1))\ldots 3.2.1}a^n(ax+b)^{m-n}$$

Hence, $\boxed{y_n = \frac{m!}{(m-n)!}a^n(ax+b)^{m-n}}$.

3. **n^{th} order derivatives of e^{mx}**

Here, $y = e^{mx}$ differentiating w.r.t. x,

$y_1 = me^{mx}$

$y_2 = m^2e^{mx}$

$y_3 = m^3e^{mx}$

... ...

Thus, $\boxed{y_n = m^n e^{mx}}$.

4. **n^{th} order derivatives of a^{mx}**

Let, $y = a^{mx}$ differentiating w.r.t. x,

$y_1 = ma^{mx}(\log a)$

$y_2 = m^2a^{mx}(\log a)^2$

$y_3 = m^3a^{mx}(\log a)^3$

... ...

Hence, $\boxed{y_n = m^n a^{mx}(\log a)^n}$.

5. **n^{th} order derivatives of** $\log(ax + b)$

In this, $y = \log(ax + b)$ differentiating w.r.t. x,

$$y_1 = \frac{a}{ax+b}$$

$$y_2 = \frac{(-1)a^2}{(ax+b)^2}$$

$$y_3 = \frac{(-1)(-2)a^3}{(ax+b)^3}$$

... ...

Hence, $\boxed{y_n = \dfrac{(-1)(-2)\ldots(-(n-1))a^n}{(ax+b)^n} = \dfrac{(-1)^{n-1}(n-1)!a^n}{(ax+b)^n}}$.

6. **n^{th} order derivatives of** $\dfrac{1}{ax+b}$

Let, $y = \dfrac{1}{ax+b}$ differentiating w.r.t. x,

$$y_1 = \frac{(-1)a}{(ax+b)^2}$$

$$y_2 = \frac{(-1)(-2)a^2}{(ax+b)^3}$$

$$y_3 = \frac{(-1)(-2)(-3)a^3}{(ax+b)^4}$$

... ...

Thus, $\boxed{y_n = \dfrac{(-1)(-2)\ldots(-n)a^n}{(ax+b)^{n+1}} = \dfrac{(-1)^n n!a^n}{(ax+b)^{n+1}}}$.

Similarly for $y = \dfrac{1}{(ax+b)^2} \Rightarrow y_n = \dfrac{(-1)^n(n+1)!a^n}{(ax+b)^{n+2}}$.

7. **n^{th} order derivatives of sin(ax + b)**

Here, $y = \sin(ax + b)$ differentiating w.r.t. x,

$$y_1 = a\cos(ax+b) = a\sin\left(ax+b+\frac{\pi}{2}\right)$$

$$y_2 = a^2\cos\left(ax+b+\frac{\pi}{2}\right) = a^2\sin\left(ax+b+\frac{2\pi}{2}\right)$$

$$y_3 = a^3\cos\left(ax+b+\frac{2\pi}{2}\right) = a^3\sin\left(ax+b+\frac{3\pi}{2}\right)$$

... ...

Thus, $\boxed{y_n = a^n\sin\left(ax+b+\frac{n\pi}{2}\right)}$.

Similarly, the reader can derive the formula for $y = \cos(ax + b)$ *i.e.*, $y_n = a^n\cos\left(ax+b+\frac{n\pi}{2}\right)$.

8. **n^{th} order derivatives of e^{ax} cos(ax + b)**

Let $y = e^{ax}\cos(ax+b)$ differentiating w.r.t. x,

$$y_1 = ae^{ax}\cos(bx+c) - be^{ax}\sin(bx+c)$$

Let $a = r\cos\theta$ and $b = r\sin\theta \Rightarrow r = \sqrt{a^2+b^2}$ and $\theta = \tan^{-1}\frac{b}{a}$

Then y_1 can be rewritten as,

$$y_1 = r\cos\theta e^{ax}\cos(bx+c) - r\sin\theta e^{ax}\sin(bx+c)$$

$$= re^{ax}(\cos\theta\cos(bx+c) - \sin\theta\sin(bx+c))$$

$$= re^{ax}\cos(bx+c+\theta)$$

$$y_2 = r^2e^{ax}\cos(bx+c+2\theta)$$

$$y_3 = r^3e^{ax}\cos(bx+c+3\theta)$$

... ...

Thus, $\boxed{y_n = r^n e^{ax}\cos(bx+c+n\theta)}$.

Similarly, $y = e^{ax}\sin(bx+c)$ *i.e.*, $y_n = r^n e^{ax}\sin(bx+c+n\theta)$, where $r = \sqrt{a^2+b^2}$ and $\theta = \tan^{-1}\frac{b}{a}$.

2.5 n^{th}–ORDER DERIVATIVES OF RATIONAL FUNCTION

To find n^{th} derivative of a rational function, firstly we have to see that the denominator consists of linear factors and then use n^{th} derivative of the algebraic function. If it does not transform into linear factors then one has to use De Moivre's theorem.

EXAMPLE 2.12 If $y = \dfrac{x^2}{(x-1)(x^2+x-2)}$ then find y_n.

$$\text{Here, } y = \frac{x^2}{(x-1)(x^2+x-2)} = \frac{x^2}{(x-1)(x-1)(x+2)} = \frac{x^2}{(x-1)^2(x+2)}.$$

$$\text{Let} \quad y = \frac{x^2}{(x-1)^2(x+2)} = \frac{A}{(x-1)} + \frac{B}{(x-1)^2} + \frac{C}{(x+2)}.$$

As we know the process of this partial fraction,

$$\therefore \qquad \frac{x^2}{(x-1)^2(x+2)} = \frac{A(x-1)(x+2) + B(x+2) + C(x-1)^2}{(x-1)^2(x+2)}.$$

$$\therefore \qquad x^2 = A(x-1)(x+2) + B(x+2) + C(x-1)^2.$$

By taking $x = 1$ in the above equation we have, $B = 1/3$.

By taking $x = -2$ in the above equation we have, $C = 4/9$.

By equating the coefficient of x^2 we have, $A + C = 1 \Rightarrow A = 5/9$.

$$\text{Hence, } y = \frac{x^2}{(x-1)^2(x+2)} = \frac{5/9}{(x-1)} + \frac{1/3}{(x-1)^2} + \frac{4/9}{(x+2)}.$$

Now, by using the formula of nth order derivative of $\dfrac{1}{ax+b}$ (defined in section 2.4(6)) we can write y_n as,

$$y_n = \frac{5}{9}\left(\frac{(-1)^n n!}{(x-1)^{n+1}}\right) + \frac{1}{3}\left(\frac{(-1)^n (n+1)!}{(x-1)^{n+2}}\right) + \frac{4}{9}\left(\frac{(-1)^n n!}{(x+2)^{n+1}}\right)$$

2.6 n^{th}–ORDER DERIVATIVES OF PRODUCTS OF THE SINE AND COSINE FUNCTIONS

EXAMPLE 2.13 If $y = \sin 6x \cos 4x$ then find y_n.

In this example, $y = \sin 6x \cos 4x$.

One can use the formula $\sin \mathrm{A} + \sin \mathrm{B} = 2\sin\left(\dfrac{A+B}{2}\right)\cos\left(\dfrac{A-B}{2}\right)$.

Thus, $y = \dfrac{\sin 10x + \cos 2x}{2}$.

From section 2.4(7), $y_n = \dfrac{10^n \sin\left(10x + \dfrac{n\pi}{2}\right) + 2^n \cos\left(2x + \dfrac{n\pi}{2}\right)}{2}$.

EXAMPLE 2.14 If $y = \sin^2 x \cos^2 x$ then find y_n.
Here,

$$y = \sin^2 x \cos^3 x = \frac{1}{4}(2\sin x \cos x)^2 \cos x = \frac{1}{4}\sin^2 2x \cos x = \frac{1}{8}(1 - \cos 4x)\cos x$$

$$\therefore \quad y = \frac{1}{8}\cos x - \frac{1}{8}\cos 4x \cos x$$

$$\therefore \quad y = \frac{1}{8}\cos x - \frac{1}{16}(\cos 5x + \cos 3x)$$

$$\therefore \quad y = \frac{1}{16}(2\cos x - \cos 5x - \cos 3x)$$

Hence, by using n^{th} order derivative of the standard function

$$\therefore \quad y_n = \frac{1}{16}\left(2\cos\left(x + \frac{n\pi}{2}\right) - 5^n \cos\left(5x + \frac{n\pi}{2}\right) - 3^n \cos\left(3x + \frac{n\pi}{2}\right)\right).$$

2.7 THE POWERS OF SINE AND COSINE FUNCTION

EXAMPLE 2.15 Find n^{th} order derivative of $\sin^4 x$.

Let $y = \sin^4 x = (\sin^2 x)^2$.

$$\therefore \quad y = \frac{1}{4}(1 - \cos 2x)^2$$

$$\therefore \quad y = \frac{1}{4}(1 - 2\cos 2x + \cos^2 2x)$$

$$\therefore \quad y = \frac{1}{4}\left(1 - 2\cos 2x + \frac{1}{2}(1 + \cos 4x)\right)$$

$$\therefore \quad y = \frac{1}{4}\left(1 - 2\cos 2x + \frac{1}{2} + \frac{\cos 4x}{2}\right)$$

$$\therefore \quad y = \frac{3}{8} - \frac{\cos 2x}{2} + \frac{\cos 4x}{8}$$

Hence, $y_n = -\frac{1}{2}2^n \cos\left(2x + \frac{n\pi}{2}\right) + \frac{1}{8}4^n \cos\left(4x + \frac{n\pi}{2}\right)$.

2.8 LEIBNITZ THEOREM

This theorem defines the formula for n^{th} order of the product of two functions. For n-times differentiable functions u and v, product will also be n-times differentiable, and that n^{th} derivative is given by, $(uv)_n = \sum_{i=0}^{n} {}^nC_i u_{n-i} v_i$.

Proof: Let u and v are two functions of x and $y = uv$.

Differentiating y w.r.t. x we have,

$$y_1 = u_1 v + u v_1$$

Converting into ${}^nC_r$ form the above equation will be, $y_1 = {}^1C_0 u_1 v + {}^1C_1 u v_1$ implies Leibnitz theorem is true for $n = 1$.

Differentiating y_1 w.r.t. x we get,

$$y_2 = (u_2 v + u_1 v_1) + (u v_2 + u_1 v_1) = u_2 v + 2u_1 v_1 + u v_2$$

$\therefore\ y_2 = {}^2C_0 u_2 v + {}^2C_1 u_1 v_1 + {}^2C_2 u v_2$ implies Leibnitz theorem is true for $n = 2$.

Differentiating y_2 w.r.t. x we get,

$$\begin{aligned} y_3 &= (u_3 v + u_2 v_1) + 2(u_2 v_1 + u_1 v_2) + (u_1 v_2 + u v_3) \\ &= u_3 v + 3u_2 v_1 + 3u_1 v_2 + u v_3 \end{aligned}$$

$\therefore\ y_3 = {}^3C_0 u_3 v + {}^3C_1 u_2 v_1 + {}^3C_2 u_1 v_2 + {}^3C_3 u v_3$ implies Leibnitz theorem is true for $n = 3$.

Now, we assume that this theorem is true for say n,

$$\begin{aligned} y_n = {}&{}^nC_0 u_n v + {}^nC_1 u_{n-1} v_1 + {}^nC_2 u_{n-2} v_2 + \ldots + {}^nC_{r-1} u_{n-r+1} v_{r-1} \\ &+ {}^nC_r u_{n-r} v_r + \ldots + {}^nC_{n-1} u_1 v_{n-1} + {}^nC_n u v_n \end{aligned}$$

Differentiating y_n w.r.t. x we get,

$$\begin{aligned} y_{n+1} = {}&{}^nC_0 (u_{n+1} v + u_n v_1) + {}^nC_1 (u_n v_1 + u_{n-1} v_2) + {}^nC_2 (u_{n-1} v_2 + u_{n-2} v_3) + \ldots \\ &+ {}^nC_{r-1} (u_{n-r+2} v_{r-1} + u_{n-r+1} v_r) + {}^nC_r (u_{n-r+1} v_r + u_{n-r} v_{r+1}) + \ldots \\ &+ {}^nC_{n-1} (u_2 v_{n-1} + u_1 v_n) + {}^nC_n (u_1 v_n + u v_{n+1}) \end{aligned}$$

$$\begin{aligned} \therefore\quad y_{n+1} = {}&{}^nC_0 u_{n+1} v + ({}^nC_0 + {}^nC_1) u_n v_1 + ({}^nC_1 + {}^nC_2) u_{n-1} v_2 + \ldots \\ &+ ({}^nC_{r-1} + {}^nC_r) u_{n-r+1} v_r + \ldots + ({}^nC_{n-1} + {}^nC_n) u_1 v_n + {}^nC_n u v_{n+1} \end{aligned}$$

(As we know, ${}^nC_{r-1} + {}^nC_r = {}^{n+1}C_r \Rightarrow {}^nC_0 + {}^nC_1 = {}^{n+1}C_1, {}^nC_1 + {}^nC_2 = {}^{n+1}C_2, \ldots$)

$$\therefore \quad y_{n+1} = {}^{n+1}C_0 u_{n+1} v + {}^{n+1}C_1 u_n v_1 + {}^{n+1}C_2 u_{n-1} v_2 + \ldots$$
$$+ {}^{n+1}C_r u_{n-r+1} v_r + \ldots + {}^{n+1}C_n u_1 v_n + {}^{n+1}C_{n+1} u v_{n+1}$$

Thus, Leibnitz's theorem is true for $n + 1$.

Hence, by mathematical induction, Leibnitz's theorem is true for every $n \in N$.

$$(uv)_n = {}^nC_0 u_n v + {}^nC_1 u_{n-1} v_1 + {}^nC_2 u_{n-2} v_2 + \ldots + {}^nC_r u_{n-r} v_r + \ldots + {}^nC_n u v_n$$
$$= \sum_{i=0}^{n} {}^nC_i u_{n-i} v_i$$

EXAMPLE 2.16 Find n^{th} order derivative of $x^2 \log x$, $n \geq 3$.

From Leibnitz's theorem,

$$(uv)_n = {}^nC_0 u_n v + {}^nC_1 u_{n-1} v_1 + {}^nC_2 u_{n-2} v_2 + \ldots + {}^nC_r u_{n-r} v_r + \ldots + {}^nC_n u v_n$$

Let $u = \log x$ and $v = x^2$

$$u_1 = \frac{1}{x},\ u_2 = -\frac{1}{x^2},\ u_3 = \frac{2}{x^3},\ \ldots,\ u_n = \frac{(-1)^{n-1}(n-1)!}{x^n}$$

$$v_1 = 2x,\ v_2 = 2,\ v_3 = 0,\ v_4 = 0,\ \ldots,\ v_n = 0.$$

Substituting in the formula we have,

$$(x^2 \log x)_n = \frac{(-1)^{n-1}(n-1)!}{x^n} x^2 + {}^nC_1 \frac{(-1)^{n-2}(n-2)!}{x^{n-1}}(2x)$$
$$+ {}^nC_2 \frac{(-1)^{n-3}(n-3)!}{x^{n-2}}(2) + 0 + \ldots$$

$$= \frac{(-1)^{n-3}}{x^{n-2}}[(-1)^2(n-1)! + {}^nC_1(-1)(n-2)!(2) + {}^nC_2(n-3)!(2)]$$

$$= \frac{(-1)^{n-3}}{x^{n-2}}\left[(n-1)(n-2)(n-3)! - n(n-2)(n-3)!(2) + \frac{n(n-1)}{2}(n-3)!(2)\right]$$

$$= \frac{(-1)^{n-3}(n-3)!}{x^{n-2}}[(n-1)(n-2) - 2n(n-2) + n(n-1)]$$

$$= \frac{(-1)^{n-3}(n-3)!}{x^{n-2}}\left[(n^2 - 3n + 2) - (2n^2 - 2n) + (n^2 - n)\right]$$

$$= \frac{2(-1)^{n-3}(n-3)!}{x^{n-2}}$$

EXAMPLE 2.17 Find n^{th} order derivative of $x^2 e^{3x} \sin 4x$.

Let, $u = e^{3x} \sin 4x$ and $v = x^2$.

$$u_n = e^{3x} 5^n \sin\left(4x + n\tan^{-1}\frac{4}{3}\right)$$

$$v_1 = 2x,\ v_2 = 2,\ v_3 = 0,\ v_4 = 0,\ \ldots,\ v_n = 0.$$

Now,

$$(x^2 e^{3x} \sin 4x)_n = x^2 e^{3x} 5^n \sin\left(4x + n\tan^{-1}\frac{4}{3}\right) + 2nxe^{3x} 5^{n-1} \sin\left(4x + (n-1)\tan^{-1}\frac{4}{3}\right)$$
$$+ n(n-1)e^{3x} 5^{n-2} \sin\left(4x + (n-2)\tan^{-1}\frac{4}{3}\right) + 0 \ldots$$

$$\therefore\ \left(x^2 e^{3x} \sin 4x\right)_n = e^{3x} 5^n \begin{bmatrix} x^2 \sin\left(4x + n\tan^{-1}\frac{4}{3}\right) + \frac{2nx}{5}\sin\left(4x + (n-1)\tan^{-1}\frac{4}{3}\right) \\ + \frac{n(n-1)}{25}\sin\left(4x + (n-2)\tan^{-1}\frac{4}{3}\right) \end{bmatrix}.$$

EXAMPLE 2.18 If $y = e^{m \sin^{-1}x}$ then prove that $(1 - x^2)y_{n+2} - (2n + 1) xy_{n+1} - (n^2 + m^2)y_n = 0$.

Here, $y = e^{m \sin^{-1}x}$

Differentiating w.r.t. x, we have

$$\Rightarrow y_1 = \frac{m}{\sqrt{1-x^2}} e^{m\sin^{-1}x} = \frac{m}{\sqrt{1-x^2}} y$$

$$\Rightarrow (1-x^2)y_1^2 = m^2 y^2 \tag{1}$$

Differentiating equation (1) w.r.t. x, we have

$$\Rightarrow (1-x^2)2y_1y_2 + y_1^2(-2x) = m^2 2yy_1$$

$$\Rightarrow (1-x^2)y_2 - xy_1 - m^2 y = 0 \tag{2}$$

Now, after n times differentiating equation (3) we get,

$$((1-x^2)y_{n+2} + {}^nC_1 y_{n+1}(-2x) + {}^nC_2 y_n(-2)) - (y_{n+1}(x) + {}^nC_1 y_n(1)) - m^2 y_n = 0$$

$$\therefore \quad (1-x^2)y_{n+2} - 2xny_{n+1} - \frac{n(n-1)}{2} y_n(2) - (y_{n+1}(x) + ny_n) - m^2 y_n = 0$$

$$\therefore \quad (1-x^2)y_{n+2} - (2n+1)xny_{n+1} - (n^2 - n + n + m^2)y_n = 0$$

$$\therefore \quad (1-x^2)y_{n+2} - (2n+1)xny_{n+1} - (n^2 + m^2)y_n = 0.$$

EXAMPLE 2.19 If $y^{1/m} + y^{-1/m} = 2x$ then prove that $(x^2 - 1)y_{n+2} + (2n + 1)xy_{n+1} + (n^2 - m^2)y_n = 0$.

Given, $$y^{1/m} + y^{-1/m} = 2x \quad (1)$$

Differentiating equation (1) w.r.t. x, we have

$$\therefore \quad \frac{1}{m}y^{(1/m)-1}y_1 - \frac{1}{m}y^{(-1/m)-1}y_1 = 2$$

$$\therefore \quad \frac{1}{m}\frac{y^{(1/m)}}{y}y_1 - \frac{1}{m}\frac{y^{(-1/m)}}{y}y_1 = 2$$

$$\therefore \quad \frac{1}{my}(y^{(1/m)} - y^{(-1/m)})y_1 = 2$$

$$\therefore \quad (y^{(1/m)} - y^{(-1/m)})y_1 = 2my$$

$$\therefore \quad (y^{(1/m)} - y^{(-1/m)})^2 y_1^2 = 4m^2y^2$$

$$\therefore \quad ((y^{(1/m)} + y^{(-1/m)})^2 - 4y^{(1/m)}y^{(-1/m)})y_1^2 = 4m^2y^2$$

$$\therefore \quad ((2x)^2 - 4)y_1^2 = 4m^2y^2$$

$$\therefore \quad (x^2 - 1)4y_1^2 = 4m^2y^2$$

$$\therefore \quad (x^2 - 1)y_1^2 = m^2y^2 \quad (2)$$

Differentiating equation (2) w.r.t. x, we have

$$2xy_1^2 + (x^2 - 1)2y_1y_2 = 2m^2yy_1$$

$$\therefore \quad xy_1 + (x^2 - 1)y_2 = m^2y \quad (3)$$

Now, after n times differentiating equation (3) we get,

$$((x^2 - 1)y_{n+2} + {}^nC_1y_{n+1}(2x) + {}^nC_2y_n(2)) + (y_{n+1}(x) + {}^nC_1y_n(1)) = m^2y_n$$

$$\therefore \quad (x^2 - 1)y_{n+2} + 2xny_{n+1} + \frac{n(n-1)}{2}y_n(2) - (y_{n+1}(x) + ny_n) = m^2y_n$$

$$\therefore \quad (x^2 - 1)y_{n+2} + (2n+1)xy_{n+1} + (n^2 - m^2)y_n = 0\,.$$

EXERCISES

Differentiate the following functions with respect to x by using the definition:

1. x
2. x^2
3. e^x
4. $\sin x$

Differentiate the following functions with respect to *x*:

5. $x^{3/2}+x^3\log x+3$

6. $\dfrac{e^x}{x}+2^x\sin x+\tan x$

7. $\dfrac{3x^3-2x-5}{x^3}$

8. $\dfrac{5+2x}{2+5x}+\log 3$

9. $\dfrac{1}{5}3^x\log x+\dfrac{3}{5}\sqrt{x}+x^{1/4}$

10. $\dfrac{(x+1)^3}{x^2}+\cot x$

Differentiate the following functions with respect to *x*:

11. $(x^2+3x)^4+\sqrt{\sin x}$

12. $\log(x^2+7)+\tan(\sin x^2)$

13. $e^{x^4-3x^2+1}+\sin(3x^2+1)$

14. $\log(\sec x+\tan x)+1$

15. $\sin(x^4)+\sin^4 x$

16. $\cos(xe^x)+\tan^{-1}(2x)$

17. Find $\dfrac{d^2y}{dx^2}$, if $y=\dfrac{1}{9}\cot(3x-1)$.

18. $f(u)=u^5+1$, $u=g(x)=\sqrt{x}$, find $(f\circ g)'$ when x = 1.

19. $f(u)=\cot\dfrac{\pi u}{10}$, $u=g(x)=5\sqrt{x}$, find $(f\circ g)'$ when x = 1.

20. If $f'(3)=-1$, $g'(2)=5$, $g(2)=3$ and $y=f(g(x))$, then $y'(2)$ = ?.

21. If y = sin(sin x), prove that $y_2+y_1\tan x+y\cos^2 x=0$.

22. If $y=x+\tan x$, then show that $y=\cos^2 x\dfrac{d^2y}{dx^2}-2y+2x=0$.

23. If $y^3+x^3+3ax^2=0$, then prove that $\dfrac{dy}{dx}=-\dfrac{(2ax+x^2)}{y^2}$.

24. If $y=Ae^{px}+Be^{qx}$, then prove that $y_2-(p+q)y_1+pqy=0$.

25. Find n^{th} derivative of $y=\sin ax+\cos ax$.

26. If $y=e^x\sin^2x$, find y_n.

27. Find y_n, if $y=\dfrac{1}{(x-1)^3(x-2)}$.

28. Find y_n, if $y=\dfrac{x^4}{(x-1)(x-2)}$.

29. Find n^{th} derivative of $x\log x$.

30. Find n^{th} derivative of x^2e^{ax}.

31. If $y=(\sin^{-1}x)^2$, prove that $(1-x^2)y_{n+2}-(2n+1)xy_{n+1}-n^2y_n=0$.

32. If $y=\log(x+\sqrt{1+x^2})$, prove that $(1+x^2)y_{n+2}+(2n+1)xy_{n+1}+n^2y_n=0$.

Answers

1. 1

2. $2x$

3. e^x

4. $\cos x$

5. $\frac{3}{2}\sqrt{x}+x^2(1+\log x)$

6. $\frac{e^x(x-1)}{x^2}+2^x(\cos x+\sin x\log 2)+\sec^2 x$

7. $\frac{4x+15}{x^4}$

8. $\frac{-21}{(2+5x)^2}$

9. $\frac{3^x}{5}\left(\frac{1}{x}+\log x\log 3\right)+\frac{3}{10\sqrt{x}}+\frac{1}{4x^{3/4}}$

10. $1-\operatorname{cosec}^2 x-\frac{2}{x^3}-\frac{3}{x^2}$

11. $4(2x+3)(x^2+3x)^3+\frac{\cos x}{2\sqrt{\sin x}}$

12. $2x\left(\frac{1}{x^2+7}+\sec^2(\sin x^2)\right)$

13. $(4x^3-6x)e^{x^4-3x^2+1}+6x\cos x(3x^2+1)$

14. $\sec x$

15. $4(x^3\cos x^4+\sin^3 x\cos x)$

16. $\frac{1}{1+4x^2}-e^x\sin(xe^x)(x+1)$

17. $2\operatorname{cosec}^2(3x-1)\cot(3x-1)$

18. $5/2$

19. $-\pi/4$

20. -5

25. $y=a^n(1+(-1)^n\sin 2ax)^{1/2}$

26. $y_n=\frac{e^x}{2}\left(1-16\left(2x+\frac{n\pi}{2}\right)\right)$

27. $y_n=\frac{(-1)^n n!}{(x-2)^{n+1}}-\frac{(-1)^n n!}{(x-1)^{n+1}}-\frac{(-1)^n(n+1)!}{(x-1)^{n+2}}-\frac{(-1)^n(n+2)!}{2(x-1)^{n+3}}$

28. $y_n=(-1)^n\, n!\left[\frac{16}{(x-2)^{n+1}}-\frac{1}{(x-1)^{n+1}}\right]$

29. $y_n=(-1)^{n-2}\frac{(n-2)!}{x^{n-1}}$

30. $y_n=a^{n-2}e^{ax}(a^2x^2+2nax+n(n-1))$.

3 Applications of Derivatives

In the previous Chapter, we understood computations and methods of derivatives. In this Chapter, we will focus on derivative applications. To know the applications, it is required to define critical points. As we studied the first derivative, second derivative, and higher order derivatives in the last chapter, and concept of critical point will help us to see where the function is monotonic.

We are familiar with the fact that the first derivative of a function gives a slope of the tangent line to the curve. We will see how the derivative affects on function's increasing and decreasing in nature. To answer this question, we will study the Mean value theorem. Before it, we need to understand the following terms and theorems.

3.1 CRITICAL POINTS

Let $f(x)$ be a function. We say that $x = k$ is a critical point of the function $f(x)$, if $f(k)$ exists and either of the two situations holds: $f'(k) = 0$ or $f'(k)$ does not exist.

Let's understand with some examples.

EXAMPLE 3.1 Find the critical points for the function, $f(x) = 6x^5 + 33x^4 - 30x^3 + 100$.

Ans: We go with the first derivative of the function, *i.e.*,

$$f(x) = 6x^5 + 33x^4 - 30x^3 + 100 \Rightarrow f'(x) = 30x^4 + 132x^3 - 90x^2$$

$$\Rightarrow f'(x) = 6x^2(5x^2 + 22x - 15)$$

$$\Rightarrow f'(x) = 6x^2(5x - 3)(x + 5)$$

In this example, the derivative of the function is polynomial which will exist everywhere. Let's find now critical points by taking, $f'(x) = 0 \Rightarrow 6x^2 (5x - 3)(x + 5) = 0$.

Hence, $x = 0$, $x = 3/5$ and $x = -5$ are the critical points.

EXAMPLE 3.2 Evaluate the critical points for the function, $f(x) = \sqrt[3]{x^2}(2x+3)$.

Ans: Here, $f(x) = \sqrt[3]{x^2}(2x+3) = 2x^{5/3} + 3x^{2/3}$

$$\therefore \quad f'(x) = \frac{10}{3}x^{2/3} + \frac{2}{3}x^{-(1/3)} = \frac{10}{3}x^{2/3} + \frac{2}{3x^{(1/3)}} = \frac{10x-2}{3x^{(1/3)}}$$

Note that, in this example

As at $x = \frac{1}{5}$, the derivative is zero, and

at $x = 0$, the derivative does not exist; there are two critical points *i.e.*, $x = \frac{1}{5}$ and $x = 0$.

EXAMPLE 3.3 Determine the critical points of $f(x) = 3xe^{4-x}$.

Ans: Firstly, $f(x) = 3xe^{4-x} \Rightarrow f'(x) = 3e^{4-x} - 3xe^{4-x}$

To find the critical points, $f'(x) = 0 \Rightarrow 3e^{4-x} - 3xe^{4-x} = 0$

$$\Rightarrow 3e^{4-x}(1-x) = 0.$$

This function will exist everywhere, only this becomes zero when $(1 - x) = 0$, *i.e.*, at $x = 1$.

EXAMPLE 3.4 Find the critical points for, $f(x) = 4 + x^2 \ln(4x)$.

Ans: $f(x) = 4 + x^2 \ln(4x) \Rightarrow f'(x) = 2x\ln(4x) + \frac{x^2}{4x} \times 4$

To find the critical points, $f'(x) = 0 \Rightarrow 2x\ln(4x) + x = 0$

$$\Rightarrow x(2\ln(4x) + 1) = 0$$

The derivative will zero only if,

$$2\ln(4x) + 1 = 0$$

$$\Rightarrow \ln(4x) = -\frac{1}{2}$$

$$\Rightarrow e^{\ln(4x)} = e^{-\frac{1}{2}}$$

$$\Rightarrow 4x = e^{-\frac{1}{2}}$$

$$\Rightarrow x = \frac{1}{4\sqrt{e}} \text{ is a critical point.}$$

Absolute/Global Extreme Values

Let $f(x)$ be a function in the domain D. Then,

(i) $f(x)$ has an **absolute maximum** value on D at a point k if $f(x) \leq f(k)$, $\forall x \in D$.

(ii) $f(x)$ has an **absolute minimum** value on D at a point k if $f(x) \geq f(k)$, $\forall x \in D$.

Relative/Local Extreme Values

Let $f(x)$ be a function in the domain D. Then,

(i) The function has a **local maximum** value at a point k within its domain if $f(x) \leq f(k)$, for all x lying in some open interval containing k.

(ii) The function has a **local minimum** value at a point k within its domain if $f(x) \geq f(k)$, for all x lying in some open interval containing k.

EXAMPLE 3.5 Find the local and global extreme values for the function, $f(x) = \cos x$.

Ans: This example will be easy if we sketch a graph for it.

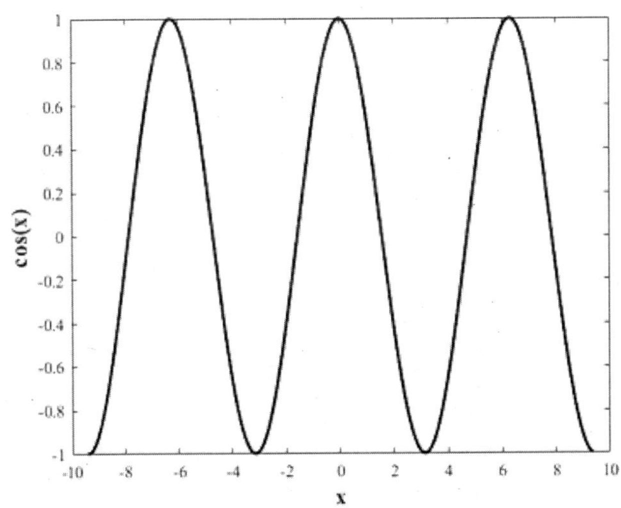

FIGURE 3.1 Graph of cos *x*.

As the particular interval is not given, $f(x) = \cos x$ has many extremum points.

It's local and global minimum $f(x) = -1$ at $x = \ldots, -3\pi, -\pi, \pi, 3\pi, \ldots.$ and $f(x) = \cos x$ has a local and global maximum $f(x) = 1$ at $x = \ldots, -4\pi, -2\pi, 0, 2\pi, 4\pi, \ldots.$

EXAMPLE 3.6 Find the local and global extreme values for the function, $f(x) = x^2$ on [–2, 2].

Ans: The graph of this function $f(x) = x^2$ in [–2, 2] is

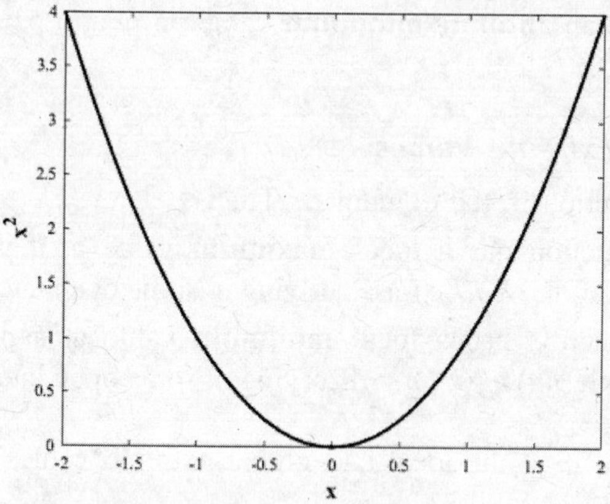

FIGURE 3.2 Graph of x^2.

From figure 3.2, we have a local and global minimum $f(x) = 0$ at $x = 0$ and global maximum $f(x) = 4$ at $x = -2$ and $x = 2$, it has no local minimum.

THEOREM 3.1 Extreme Value Theorem
If $f(x)$ is a continuous function on a closed interval $[a, b]$, then $f(x)$ attains an absolute maximum value (say) M and an absolute minimum value (say) m in $[a, b]$. That is, for any $x \in [a, b]$, $m \le f(x) \le M$.

EXAMPLE 3.7 Find the absolute extrema for the function $f(x) = 2x^3 + 3x^2 - 12x + 4$ on [–4, 2].

Ans: To find critical points of the function, we take the derivative of the function, $f(x) = 2x^3 + 3x^2 - 12x + 4$.

$$\Rightarrow f'(x) = 6x^2 + 6x - 12$$

$$\Rightarrow f'(x) = 6(x^2 + x - 2)$$

$$\Rightarrow f'(x) = 6(x+2)(x-1)$$

Now, $f'(x) = 0 \Rightarrow x = -2, x = 1$.

Function value at critical points: $f(-2) = 2(-2)^3 + 3(-2)^2 - 12(-2) + 4 = 24$

$$f(1) = 2(1)^3 + 3(1)^2 - 12(1) + 4 = 24 = -3$$

Function value at endpoints: $f(-4) = 2(-4)^3 + 3(-4)^2 - 12(-4) + 4 = -28$

$$f(2) = 2(2)^3 + 3(2)^2 - 12(2) + 4 = 8$$

From this, absolute extrema will occur at only these four points.

Hence, the absolute maximum value $f(x) = 24$ occurs at $x = -2$, the absolute minimum value $f(x) = -28$ occurs at $x = -4$.

EXAMPLE 3.8 Determine the absolute extrema for the function $f(p) = 3p(p+4)^{2/3}$ on [–5, –1].

Ans: We first take derivative for finding critical points:

$$f(p) = 3p(p+4)^{2/3}$$

$$\therefore \quad f'(p) = 3(p+4)^{2/3} + 3p\left(\frac{2}{3}\right)(p+4)^{-1/3}$$

$$\therefore \quad f'(p) = 3(p+4)^{2/3} + \frac{2p}{(p+4)^{1/3}}$$

$$\therefore \quad f'(p) = \frac{3(p+4)+2p}{(p+4)^{1/3}} = \frac{5p+12}{(p+4)^{1/3}}$$

Now, at $p = -12/5$ the derivative is zero.
and at $p = -4$ the derivative doesn't exist. So, these are two critical points.

Function value at critical points: $f(-4) = 0$ and $f\left(-\frac{12}{5}\right) = -9.8495$

at endpoints: $f(-5) = -15$ and $f(-1) = -6.2403$

First Derivative Test

Let $f(x)$ be a function, having $x = k$ is a critical point then,

(i) If $f'(x) > 0$ for all $x < k$ and if $f'(x) < 0$ for all $x > k$, then $x = k$ is a relative maximum.

(ii) If $f'(x) < 0$ for all $x > k$ and $f'(x) > 0$ for all $x < k$, then $x = k$ is a relative minimum.

(iii) If $f'(x)$ has any other situation then, the critical point is neither a relative maximum nor a minimum.

[**Note**: If f has a local extremum value at an interior point c in its domain and f' is defined at c then $f'(c) = 0$.]

EXAMPLE 3.9 Determine the intervals for the function $f(x) = x\sqrt[3]{x^2-4}$ where the function is increasing and decreasing.

Ans: To find critical points, we take the derivative of $f(x) = x\sqrt[3]{x^2-4}$

$$f'(x) = \sqrt[3]{x^2-4} + x\left(\frac{1}{3}2x\right)(x^2-4)^{-2/3}$$

$$\Rightarrow f'(x) = \sqrt[3]{x^2 - 4} + \frac{2x^2}{3(x^2 - 4)^{2/3}}$$

$$\Rightarrow f'(x) = \frac{3(x^2 - 4) + 2x^2}{3(x^2 - 4)^{2/3}} = \frac{5x^2 - 12}{3(x^2 - 4)^{2/3}}$$

Therefore, here we have critical points:

$x = \pm\sqrt{\frac{12}{5}} = \pm 1.5492$ here, the derivative is zero.

$x = \pm 2$ here, the derivative doesn't exist.

After finding these critical points, we will check the function value of $f'(x)$ at points near to critical points.

We have, $f'(x) = \sqrt[3]{x^2 - 4} + \dfrac{2x^2}{3(x^2 - 4)^{2/3}}$.

Putting the value at $x = \pm 3$ onwards, $f'(3) = 3.76$ and $f'(-3) = 3.76$ implies $f'(x) > 0$.

at $x = \pm 1.6$, $f'(\pm 1.6) = -2.47$ implies $f'(x) < 0$.

at $x = \pm 1.7$ onwards, $f'(\pm 1.7) = 0.76$ implies $f'(x) > 0$.

at $x = 0$, $f'(0) = -1.59$ implies $f'(x) < 0$.

Thus, the function is:

increasing interval: $-\infty < x < -2,\ -2 < x < -\sqrt{\frac{12}{5}},\ \sqrt{\frac{12}{5}} < x < 2,\ 2 < x < \infty$

decreasing interval: $-\sqrt{\frac{12}{5}} < x < \sqrt{\frac{12}{5}}$.

We can observe that at $x = \pm 2$ the function has neither maximum nor minimum.

At $x = \sqrt{\frac{12}{5}}$, the function has a relative minimum, and $x = -\sqrt{\frac{12}{5}}$, the function has a relative maximum.

EXAMPLE 3.10 Let the function $f(x) = x^3$ be defined. Find relative extremum.

Ans: Here, $f(x) = x^3 \Rightarrow f'(x) = 3x^2$

Now, $f'(x) = 0 \Rightarrow x = 0$.

Let us check now points where the function is increasing and decreasing.

At $x = 1 \Rightarrow f'(1) = 3(1)^2 = 3, f'(1) > 0$,

$x = -1 \Rightarrow f'(-1) = 3(-1)^2 = 3, f'(-1) > 0$

As the slope of the function will be positive, this function is only increasing in nature. The function is not changing in nature, so one cannot find extremum.

Increasing interval: $x \in (-\infty, 0) \cup (0, \infty)$

Consequently, there is no relative extremum.

3.2 ROLLE'S THEOREM

If $f(x)$ is a function that satisfies the following conditions:

1. $f(x)$ is continuous on $[a, b]$.
2. $f(x)$ is differentiable on (a, b).
3. $f(a) = f(b)$.

Then there is at least a number c in (a, b) at which $f'(c) = 0$.

Proof: Here, $f(x)$ is a continuous function on $[a, b]$, hence from Theorem 1 function will attain maximum and minimum values somewhere in $[a, b]$. These extreme values occur at

(i) the interior point where f' is zero,

(ii) the interior point where f' does not exist,

(iii) the endpoints i.e., a and b.

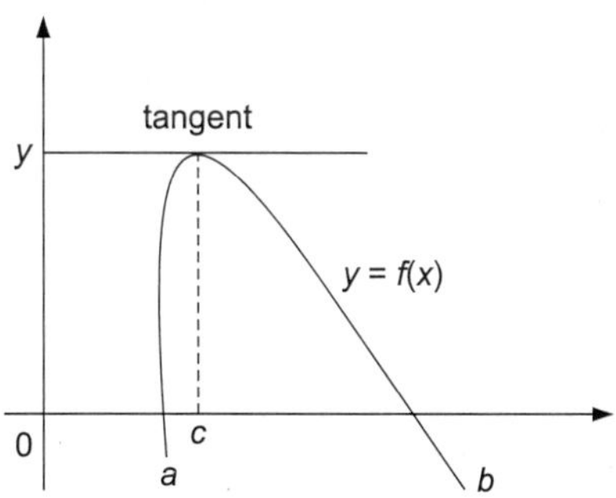

FIGURE 3.3 Graph of function $y = f(x)$.

From the first derivative test, if the extremum point occurs at the point c between a and b then we can say that $f'(c)$ exists and as it is the critical point $f'(c) = 0$.

If maximum and minimum values occur at endpoints then, as $f(a) = f(b)$ it will be a constant function with $f(x) = f(a) = f(b)$ for every $x \in [a, b]$. Therefore, $f'(x) = 0$. and point c will be anywhere between a and b.

EXAMPLE 3.11 If $f(x) = x^2 + 2x$, find the values of c in the interval $[-2, 0]$ such that $f'(c) = 0$.

Ans: As the function $f(x) = x^2 + 2x$ is quadratic polynomial, it is continuous in [–2, 0]. It is also differentiable in (–2, 0).

Function value at endpoints:

$$f(-2) = (-2)^2 + 2(-2) = 0$$

$$f(0) = (0)^2 + 2(0) = 0 \text{ implies } f(-2) = f(0).$$

Thus, the function satisfies all three conditions hence we can apply Rolle's theorem.

Therefore, there exists $c \in [-2, 0]$ such that $f'(c) = 0$.

$$f'(x) = 2x + 2$$

Now, $f'(c) = 0 \Rightarrow 2c + 2 = 0$

$$\Rightarrow c = -1.$$

EXAMPLE 3.12 Find the value of c using Rolle's theorem the function $f(x) = \cos\left(\frac{x}{2}\right)$ on $[\pi, 3\pi]$.

Ans: Here, the given function is $f(x) = \cos\left(\frac{x}{2}\right)$.

$$f(\pi) = \cos\left(\frac{\pi}{2}\right) = 0 \text{ and } f(3\pi) = \cos\left(\frac{3\pi}{2}\right) = 0.$$

Thus, as $f(\pi) = f(3\pi)$, there exists $c \in [\pi, 3\pi]$ such that $f'(c) = 0$.

$$f'(x) = -\frac{1}{2}\left(\sin\frac{x}{2}\right)$$

Now, $f'(c) = -\frac{1}{2}\left(\sin\frac{c}{2}\right) = 0$

$$\Rightarrow \sin\frac{c}{2} = 0$$

$$\Rightarrow \frac{c}{2} = n\pi$$

$\Rightarrow c = 2n\pi$, where n is an integer.

We want $c \in [\pi, 3\pi]$, implies $c = 2\pi$.

EXAMPLE 3.13 Find the interval $[a, b]$ for the function $f(x) = x^4 + x^2 - 2$ which satisfies Rolle's theorem, if $a = -1$.

Ans: Given the function

$$f(x) = x^4 + x^2 - 2 = (x^2 + 2)(x^2 - 1) = (x^2 + 2)(x - 1)(x + 1)$$

As the function has two zeros and it satisfies Rolle's theorem, the required interval will be [–1, 1] implies $b = 1$.

3.3 THE MEAN VALUE THEOREM

Suppose $y = f(x)$ is a continuous function on a closed interval $[a, b]$ and a differentiable function on an open interval (a, b). Then there is a number c in (a, b) such that,

$$f'(c) = \frac{f(b) - f(a)}{b - a}$$

Proof: Suppose the graph of $y = f(x)$ is,

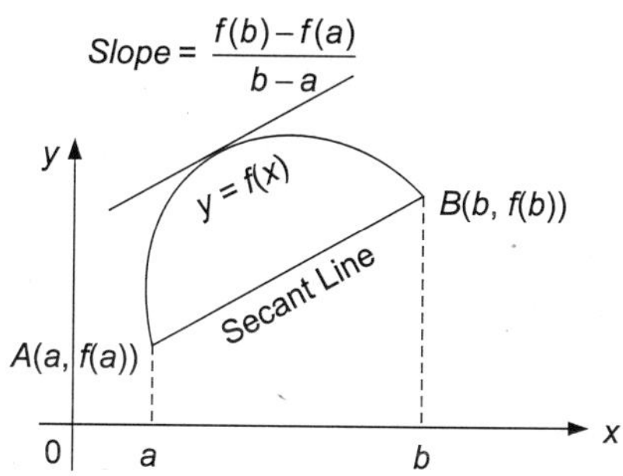

FIGURE 3.4 Graph for slope of chord AB.

From the figure, the equation of the line through the two points $A(a, f(a))$ and $B(b, f(b))$ is,

$$y = f(x) = f(a) + \frac{f(b) - f(a)}{b - a}(x - a) \tag{3.1}$$

Let us now define the new equation as the difference between $y = f(x)$ and the equation of the secant line,

$$g(x) = f(x) - \left(f(a) + \frac{f(b) - f(a)}{b - a}(x - a)\right) = f(x) - f(a) - \frac{f(b) - f(a)}{b - a}(x - a) \tag{3.2}$$

Note that, as the sum of $f(x)$ which is continuous on $[a, b]$ and differentiable on (a, b) and a linear polynomial; the function $g(x)$ is continuous on $[a, b]$ and differentiable on (a, b).

Hence, by taking the derivative of equation (2) we have,

$$g'(x) = f'(x) - \frac{f(b) - f(a)}{b - a} \tag{3.3}$$

and also, the value of the function $g(x)$ at endpoints (a, b) we have,

$$g(a) = f(a) - f(a) - \frac{f(b)-f(a)}{b-a}(a-a) = f(a) - f(a) = 0$$

$$g(b) = f(b) - f(a) - \frac{f(b)-f(a)}{b-a}(b-a) = f(b) - f(a) - (f(b) - f(a)) = 0$$

Thus, $g(x)$ satisfies three conditions of Rolle's Theorem.

Therefore, a number c in (a, b) at which $f'(c) = 0$, *i.e.*,

$$g'(c) = f'(c) - \frac{f(b)-f(a)}{b-a} = 0$$

$$\therefore \qquad f'(c) = \frac{f(b)-f(a)}{b-a}.$$

Hence the proof.

EXAMPLE 3.14 Find the value c which satisfies the Mean Value Theorem for function $f(x) = x^3 + 2x^2 - x$ on $[-1, 2]$.

Ans: As the given function is polynomial, the function is continuous and differentiable on the interval.

$$\text{Now, } f(x) = x^3 + 2x^2 - x \Rightarrow f'(x) = 3x^2 + 4x - 1$$

Given function satisfies Mean Value Theorem,

$$\Rightarrow f'(c) = \frac{f(2)-f(-1)}{2-(-1)}$$

$$f(2) = (2)^3 + 2(2)^2 - 2 = 14$$

$$f(-1) = (-1)^3 + 2(-1)^2 + 1 = 2$$

Now putting values in $f'(c)$,

$$\Rightarrow 3c^2 + 4c - 1 = \frac{14-2}{3} = 4$$

$$\Rightarrow 3c^2 + 4c - 1 = 4$$

$$\Rightarrow 3c^2 + 4c - 5 = 0$$

$$\Rightarrow c = \frac{-4 \pm \sqrt{76}}{6}$$

Here, c possess two values 0.7863 and –2.1196.

But the value of $c \in [-1, 2]$ is 0.7863 implies $c = 0.7863$.

EXAMPLE 3.15 Determine the value of c when $f(x) = x + 3\cos x$ on $[-\pi, \pi]$ using the Mean Value Theorem.

Ans: Given $f(x) = x + 3\cos x$

$$f(\pi) = \pi + 3\cos\pi = \pi + 3(-1) = \pi - 3$$

$$f(-\pi) = -\pi + 3\cos(-\pi) = -\pi + 3(-1) = -\pi - 3$$

Taking derivative, $f'(x) = 1 - 3\sin x$

By applying Mean Value Theorem, we have

$$f'(c) = \frac{f(\pi) - f(-\pi)}{\pi - (-\pi)} \Rightarrow 1 - 3\sin c = \frac{\pi - 3 - (-\pi - 3)}{2\pi}$$

$$\Rightarrow 1 - 3\sin c = \frac{2\pi}{2\pi}$$

$$\Rightarrow 1 - 3\sin c = 1$$

$$\Rightarrow -3\sin c = 0$$

$$\Rightarrow \sin c = 0$$

$$\Rightarrow c = n\pi,\ n = 0, 1, 2, \ldots$$

But $c \in [-\pi, \pi]$ implies $c = 0$.

3.4 GENERALIZED MEAN VALUE THEOREM

If $f_1(x)$ and $f_2(x)$ are two continuous functions on a closed interval $[a, b]$ and differentiable on an open interval (a, b) and have identical derivatives on that interval then $f_1(x) - f_2(x)$ is a constant.

Proof: Let's assume the difference function $g(x) = f_1(x) - f_2(x)$

$$\therefore \quad g'(x) = f_1'(x) - f_2'(x) = 0 \text{ at each point in } (a, b)$$

$$\Rightarrow g(x) = f_1(x) - f_2(x) = \text{constant.}$$

3.5 CAUCHY'S MEAN VALUE THEOREM (CMVT)

Suppose that $f(x)$ and $g(x)$ are continuous functions on a closed interval $[a, b]$ and differentiable on an open interval (a, b), with $g'(x) \neq 0$, $\forall x \in (a, b)$. Then there exists $c \in (a, b)$ such that,

$$\frac{f'(c)}{g'(c)} = \frac{f(b) - f(a)}{g(b) - g(a)}$$

Proof: Let's assume that, $\phi(x) = f(x)[g(b) - g(a)] - g(x)[f(b) - f(a)]$

Now calculating at endpoints $x = a$ and $x = b$ we get,

$$\phi(a) = f(a)[g(b) - g(a)] - g(a)[f(b) - f(a)] = f(a)g(b) - g(a)f(b)$$

and $\phi(b) = f(a)g(b) - g(a)f(b)$.

Thus, $\phi(a) = \phi(b)$ suggest that $\phi(x)$ satisfies Rolle's theorem.

Therefore, there exists $c \in (a, b)$ such that $\phi'(c) = 0$.

$$\phi'(c) = f'(c)[g(b) - g(a)] - g'(c)[f(b) - f(a)] = 0$$

$$\Rightarrow \frac{f'(c)}{g'(c)} = \frac{f(b) - f(a)}{g(b) - g(a)}$$; hence the proof.

EXAMPLE 3.16 Verify whether Cauchy's Mean Value Theorem satisfy for the function $f(x) = x^4$ and $g(x) = x^2$ in the interval [1, 2].

Ans: The functions $f(x) = x^4$ and $g(x) = x^2$.

The derivatives of functions $f'(x) = 4x^3$ and $g'(x) = 2x$

Putting these values in the CMVT formula we have,

$$\Rightarrow \frac{f'(c)}{g'(c)} = \frac{f(b) - f(a)}{g(b) - g(a)}$$

$$\Rightarrow \frac{4c^3}{2c} = \frac{b^4 - a^4}{b^2 - a^2}$$

$$\Rightarrow 2c^2 = \frac{(b^2 - a^2)(b^2 + a^2)}{b^2 - a^2}$$

$$\Rightarrow c^2 = \frac{b^2 + a^2}{2}$$

$$\Rightarrow c = \pm\sqrt{\frac{a^2 + b^2}{2}}$$

If we take the boundary value of interval [1, 2], i.e., $a = 1$ and $b = 2$.

Hence, $\Rightarrow c = \pm\sqrt{\frac{a^2 + b^2}{2}} = \pm\sqrt{\frac{1^2 + 2^2}{2}} = \pm 1.58$

The value of $c = 1.58$ which $c \in [1, 2]$ implies that functions satisfy the CMVT.

EXAMPLE 3.17 Verify CMVT for the functions $f(x) = \cos x$ and $g(x) = \sin x$ on the interval $[a, b]$.

Ans: Given $f(x) = \cos x$ and $g(x) = \sin x$

Taking derivatives $f'(x) = -\sin x$ and $g'(x) = \cos x$

Substituting in the CMVT formula we get,

$$\Rightarrow \frac{f'(c)}{g'(c)} = \frac{f(b)-f(a)}{g(b)-g(a)}$$

$$\Rightarrow -\frac{\sin c}{\cos c} = \frac{\cos b - \cos a}{\sin b - \sin a}$$

$$\Rightarrow -\tan c = \frac{-2\sin\frac{b+a}{2}\sin\frac{b-a}{2}}{2\cos\frac{b+a}{2}\sin\frac{b-a}{2}}$$

$$\Rightarrow -\tan c = -\tan\frac{a+b}{2}$$

$$\Rightarrow c = \frac{a+b}{2} + n\pi,\ n \in Z$$

If we take $n = 0$, $\Rightarrow c = \dfrac{a+b}{2}$.

Now, we need to find the interval in which the solution exists.

denominator $\neq 0$

$$\Rightarrow \sin\frac{b-a}{2} \neq 0 \text{ and } \cos\frac{a+b}{2} \neq 0.$$

$$\Rightarrow \frac{b-a}{2} \neq n\pi \text{ and } \frac{a+b}{2} \neq \frac{\pi}{2} + n\pi$$

$$\Rightarrow b-a \neq 2n\pi \text{ and } a+b \neq \pi + 2n\pi$$

$$\Rightarrow a \neq \frac{\pi}{2} + n\pi \text{ and } b \neq \frac{\pi}{2} + n\pi$$

Extended Generalized Mean Value Theorem:

Let $n \in \mathbb{N}$ and $f(x), g_1(x), g_2(x), \ldots, g_n(x)$ be $n + 1$ continuous functions on a closed interval $[a, b]$ and differentiable in an open interval (a, b), with $g_i'(x) \neq 0, \forall x \in (a, b)$ for $i = 1, 2, n$. Then there exists $c \in (a, b)$:

$$f'(c) = \frac{f(b)-f(a)}{n}\left[\sum_{i=1}^{n} \frac{g_i'(c)}{g_i(b)-g_i(a)}\right]$$

Proof: Let's assume a function $F(x)$ by:

$$F(x) = n(f(x) - f(a)) - \sum_{i=1}^{n} \frac{f(b) - f(a)}{g_i(b) - g_i(a)} (g_i(x) - g_i(a))$$

Given $g'(x) \neq 0 \Rightarrow g_i(b) - g_i(a) \neq 0$ for $i = 1, 2, 3, ..., n$.

Moreover $F(a) = F(b) = 0$.

Also, $F(x)$ is a continuous function on $[a, b]$ and $F(x)$ is a differentiable function in (a, b) as $F(x)$, $g_i(x)$ for $i = 1, 2, ..., n$ is continuous and differentiable over an interval.

Hence by Rolle's theorem, there exists $c \in (a, b)$ such that $F'(c) = 0$,

$$F'(x) = nf'(c) - \sum_{i=1}^{n} \frac{f(b) - f(a)}{g_i(b) - g_i(a)} g_i'(c) = 0$$

$$\Rightarrow f'(c) = \frac{f(b) - f(a)}{n} \left[\sum_{i=1}^{n} \frac{g_i'(c)}{g_i(b) - g_i(a)} \right]$$ hence the proof.

3.6 MEAN VALUE THEOREM OF HIGHER DERIVATIVES

Let $n \in \mathbb{N}$ and $f(x)$, $g_i(x)$ for $i = 1, 2, ..., n$ be $n + 1$ continuous functions on a closed interval $[a, b]$ and are k times differentiable in an open interval (a, b), with $g_i^k(x) \neq 0$, $\forall x \in (a, b)$ for $i = 1, 2, ..., n$. Then there exists $c \in (a, b)$ such that,

$$f^k(c) = \frac{f^{(k-1)}(b) - f^{(k-1)}(a)}{n} \left[\sum_{i=1}^{n} \frac{g_i^k(c)}{g_i^{(k-1)}(b) - g_i^{(k-1)}(a)} \right]$$

Proof: Suppose the function $\phi(x) = f^{(k-1)}(x)$ and $\psi_i(x) = g_i^{(k-1)}(x)$ for $i = 1, 2, ..., n$.

It follows that, $\psi_i'(x) = g_i^k(x) \neq 0$.

Moreover $\phi(x)$ and $\psi_i(x)$ are continuous on $[a, b]$ and differentiable on (a, b).

Hence extended mean value theorem, there exists $c \in (a, b)$ such that,

$$\Rightarrow \phi'(c) = \frac{\phi(b) - \phi(a)}{n} \left[\sum_{i=1}^{n} \frac{\psi_i'(c)}{\psi_i(b) - \psi_i(a)} \right]$$

$$\Rightarrow f^k(c) = \frac{f^{(k-1)}(b) - f^{(k-1)}(a)}{n} \left[\sum_{i=1}^{n} \frac{g_i^k(c)}{g_i^{(k-1)}(b) - g_i^{(k-1)}(a)} \right],$$

hence the proof.

3.7 L'HOSPITAL'S RULE FOR THE INTERMEDIATE FORMS

Going back to the concept of limit and let's see some of the limits like,

$$\lim_{x\to\infty}\frac{2x+3}{x-6},\ \lim_{x\to a}\frac{x^2-a^2}{x-a},\ \lim_{x\to 0}\frac{\tan x}{x},\ \lim_{x\to 0}\frac{1}{x},\ \lim_{x\to\infty}\frac{e^x}{x}$$

If we plug the limit, then they are like ∞/∞ or 0/0 forms which are known as **Indeterminate forms.** One can solve the following examples,

$$\lim_{x\to\infty}\frac{2x+3}{x-6}=\lim_{x\to\infty}\frac{x\left(2+\frac{3}{x}\right)}{x\left(1-\frac{6}{x}\right)}=\frac{2+0}{1-0}=2 \text{ and}$$

$$\lim_{x\to a}\frac{x^2-a^2}{x-a}=\lim_{x\to a}\frac{(x-a)(x+a)}{x-a}=2a$$

But the rest of all expressions seems undefined or determined or the limit does not exist, it brings contradiction to our perception.

Thus in the quotient $\lim_{x\to a}\frac{f(x)}{g(x)}$, if both $f(x)\to 0$ and $g(x)\to 0$ as $x\to a$ then it is an indeterminate form of type $\frac{0}{0}$. Similarly, if both $f(x)\to\pm\infty$ and $g(x)\to\pm\infty$ as $x\to a$ then it is an indeterminate form of type $\frac{\infty}{\infty}$.

For this **L'Hospital's** or **L'Hôpital's Rule** comes into the picture. Let's see its definition and how it works.

Statement: Let $f(x)$ and $g(x)$ be differentiable and continuous in an interval I containing $a\in\mathbb{R}$ and $g'(a)\neq 0$ on I for $x\neq a$. If $\lim_{x\to a}\frac{f(x)}{g(x)}=\frac{0}{0}$ or $\frac{\infty}{\infty}$ form then in these cases we have, $\lim_{x\to a}\frac{f(x)}{g(x)}=\lim_{x\to a}\frac{f'(x)}{g'(x)}$.

Proof: Here, $f(x)$, $g(x)$, $f'(x)$ and $g'(x)$ are continuous in the interval I containing $a\in\mathbb{R}$.

$$\lim_{x\to a}f(x)=0=\lim_{x\to a}g(x) \text{ implies } f(a)=0=g(a).$$

$$\lim_{x\to a}\frac{f(x)}{g(x)}=\lim_{x\to a}\frac{f(x)-f(a)}{g(x)-g(a)} \qquad (\because f(a)=0=g(a))$$

$$=\lim_{x\to a}\frac{f(x)-f(a)}{g(x)-g(a)}$$

$$= \lim_{x \to a} \frac{\dfrac{f(x) - f(a)}{x - a}}{\dfrac{g(x) - g(a)}{x - a}}$$

$$= \frac{f'(a)}{g'(a)}$$

$$= \lim_{x \to a} \frac{f'(x)}{g'(x)}.$$

$\dfrac{0}{0}$ Indeterminate Form

EXAMPLE 3.18 Evaluate $\lim\limits_{x \to 1} \dfrac{\sin(\pi x)}{\ln x}$.

Ans: As $x \to 1$, $\sin(\pi x) \to 0$ and $\ln x \to 0$.

Therefore, by applying L'Hospital's Rule we have,

$$\lim_{x \to 1} \frac{\sin(\pi x)}{\ln x} = \lim_{x \to 1} \frac{\pi \cos(\pi x)}{1/x} = \frac{\pi \cdot \cos \pi}{1} = -\pi.$$

EXAMPLE 3.19 Evaluate $\lim\limits_{x \to \infty} \dfrac{e^{1/x} - 1}{1/x}$.

Ans: As $x \to \infty, e^{1/x} - 1 \to 0$ and $1/x \to 0$.

By L'Hospital's Rule, we obtain,

$$\therefore \quad \lim_{x \to \infty} \frac{e^{1/x} - 1}{1/x} = \lim_{x \to \infty} \frac{e^{1/x}(-1/x^2)}{(-1/x^2)} = \lim_{x \to \infty} e^{1/x} = e^0 = 1..$$

$\dfrac{\infty}{\infty}$ Indeterminate Form

EXAMPLE 3.20 Evaluate $\lim\limits_{x \to 0^+} \dfrac{\ln x}{\cot x}$.

Ans: As $x \to 0^+$, $\ln x \to \infty$ and $\cot x \to \infty$.

Hence by L'Hospital's Rule, we get,

$$\lim_{x \to 0^+} \frac{\ln x}{\cot x} = \lim_{x \to 0^+} \frac{1/x}{-\operatorname{cosec}^2 x} = \lim_{x \to 0^+} \frac{\sin^2 x}{-x}$$

which is $\dfrac{0}{0}$ form, again applying L'Hospital's Rule we have,

$$\lim_{x \to 0^+} \frac{\sin^2 x}{-x} = \lim_{x \to 0^+} \frac{2 \sin x \cos x}{-1} = \frac{0}{-1} = 0.$$

$0 \cdot \infty$ Indeterminate Form

EXAMPLE 3.21 Evaluate $\lim_{x \to 0} x \cot x$.

Ans: For applying L'Hospital's Rule we convert the function into fractions,

$$\lim_{x \to 0} x \cot x = \lim_{x \to 0} \frac{x}{\tan x} = \lim_{x \to 0} \frac{1}{\sec^2 x} = \frac{1}{1} = 1$$

0^0 Indeterminate Form

EXAMPLE 3.22 Evaluate $\lim_{x \to 0^+} x^{\sin x}$.

Ans: Let $y = x^{\sin x}$

Taking log both sides,

$$\Rightarrow \ln y = \ln x^{\sin x}$$

$$\Rightarrow \ln y = \sin x \ln x$$

As $x \to 0^+$, $\sin x \to 0$ and $\ln x \to \infty$. To apply L'Hospital's Rule we convert the function into fractions,

$$\Rightarrow \ln y = \sin x \ln x = \frac{\ln x}{1/\sin x}$$

$$\Rightarrow \lim_{x \to 0^+} \sin x \ln x = \lim_{x \to 0^+} \frac{\ln x}{1/\sin x} = \lim_{x \to 0^+} \frac{\ln x}{\operatorname{cosec} x}$$

$$\Rightarrow \lim_{x \to 0^+} \sin x \ln x = \lim_{x \to 0^+} \frac{1/x}{-\operatorname{cosec} x \cot x} = \lim_{x \to 0^+} \frac{1}{-x \operatorname{cosec} x \cot x}$$

$$= \lim_{x \to 0^+} \frac{1}{-x \dfrac{1}{\sin x} \dfrac{1}{\tan x}}$$

$$\Rightarrow \lim_{x \to 0^+} \sin x \ln x = \lim_{x \to 0^+} \left(- \frac{\sin x}{x} \cdot \tan x \right)$$

$$= \left(\lim_{x \to 0^+} \frac{\sin x}{x} \right) \cdot \lim_{x \to 0^+} (-\tan x) = 1 \cdot 0 = 0$$

Hence, $\lim_{x \to 0^+} \ln y = \lim_{x \to 0^+} \sin x \ln x = 0$

$$\Rightarrow \lim_{x \to 0^+} y = \lim_{x \to 0^+} x^{\sin x} = e^0 = 1$$

∞^0 Indeterminate Form

EXAMPLE 3.23 Evaluate $\lim\limits_{x\to\infty} x^{1/x}$.

Ans: Let $y = x^{1/x}$

Taking log both sides,

$$\ln y = \frac{1}{x}\ln x = \frac{\ln x}{x}$$

$$\Rightarrow \lim_{x\to\infty} \ln y = \lim_{x\to\infty} \frac{\ln x}{x} = \lim_{x\to\infty} \frac{1/x}{1} = 0$$

Now, $\lim\limits_{x\to\infty} y = \lim\limits_{x\to\infty} x^{1/x} = \lim\limits_{x\to\infty} e^{\ln y} = e^{\lim\limits_{x\to\infty} \ln y} = e^0 = 1$.

1^∞ Indeterminate Form

EXAMPLE 3.24 Evaluate $\lim\limits_{x\to 0^+} (e^x + x)^{1/x}$.

Ans: Here, $\lim\limits_{x\to 0^+} (e^x + x)^{1/x} = \lim\limits_{x\to 0^+} e^{\ln (e^x + x)^{1/x}} = \lim\limits_{x\to 0^+} e^{\frac{\ln(e^x + x)}{x}}$

Take $\lim\limits_{x\to 0^+} y = \lim\limits_{x\to 0^+} (e^x + x)^{1/x}$

After taking log both sides we have,

$$\lim_{x\to 0^+} \ln y = \lim_{x\to 0^+} \frac{\ln(e^x + x)}{x}$$

$$\Rightarrow \lim_{x\to 0^+} \ln y = \lim_{x\to 0^+} \frac{\frac{1}{e^x + x}\cdot(e^x + 1)}{1} = \lim_{x\to 0^+} \frac{e^x + 1}{e^x + x} = \frac{1+1}{1+0} = 2$$

Hence, $\lim\limits_{x\to 0^+} y = \lim\limits_{x\to 0^+} (e^x + x)^{1/x} = e^2$.

$\infty - \infty$ Indeterminate Form

EXAMPLE 3.25 Evaluate $\lim\limits_{x\to 1^+} \left(\dfrac{1}{x-1} - \dfrac{1}{\ln x}\right)$.

Ans: This is $\infty - \infty$ form,

$$\lim_{x\to 1^+} \left(\frac{1}{x-1} - \frac{1}{\ln x}\right) = \lim_{x\to 1^+} \frac{\ln x - (x-1)}{(x-1)(\ln x)}$$

Applying L'Hospital's Rule we have,

$$\Rightarrow \lim_{x\to 1^+}\left(\frac{1}{x-1}-\frac{1}{\ln x}\right)=\lim_{x\to 1^+}\frac{\frac{1}{x}-1}{(\ln x)+(x-1)\frac{1}{x}}$$

Applying L'Hospital's Rule we have,

$$\Rightarrow \lim_{x\to 1^+}\left(\frac{1}{x-1}-\frac{1}{\ln x}\right)=\lim_{x\to 1^+}\frac{\frac{-1}{x^2}}{\frac{1}{x}+(x-1)\frac{-1}{x^2}+\frac{1}{x}}$$

$$\Rightarrow \lim_{x\to 1^+}\left(\frac{1}{x-1}-\frac{1}{\ln x}\right)=\lim_{x\to 1^+}\frac{\frac{-1}{x^2}}{\frac{1}{x}+\frac{-1}{x}+\frac{1}{x^2}+\frac{1}{x}}$$

$$\Rightarrow \lim_{x\to 1^+}\left(\frac{1}{x-1}-\frac{1}{\ln x}\right)=\lim_{x\to 1^+}\frac{\frac{-1}{x^2}}{\frac{1}{x}+\frac{1}{x^2}}=-\frac{1}{2}$$

EXERCISES

Find Critical points for the following examples:

1. x^3+3x^2-24x
2. $10te^{3-t^2}$
3. $\sin x+\cos x$ on $[\pi, 2\pi]$
4. $1-|x+2|$
5. $\dfrac{e^x}{x}$
6. $(x+2)^{2/3}(2x-1)$

Find relative extremum for the following functions:

7. $\sin x+\cos x$
8. x^3
9. $2x-3x^{2/3}$

Find absolute extremum for the following functions:

10. $\sin\left(\dfrac{x}{3}\right)+\dfrac{2x}{9}$ on $[-10, 15]$
11. $4x^3-3x^2+9x+12$ on $[-2, 1]$
12. $\ln(x^2+4x+14)$ on $[-4, 2]$

Determine the value of *c* which satisfies Rolle's theorem:

13. $x^3 - x$ on $[-1, 1]$

14. $x^2 + 8x + 14$

15. $\log\left(\dfrac{x^2 + ab}{x(a+b)}\right), x \in [a,b] - \{0\}$

16. $e^x \sin x, \forall x \in [0, \pi]$

17. The function $f(x) = x^3 - 2x^2 + 3$ satisfies Rolle's Theorem defined on $[a, b]$. If the value of $a = 0$, then $b = ?$

Determine the value of *c* which satisfy the Mean Value Theorem:

18. $3 + \sqrt{x}$ on $[0, 4]$

19. $\dfrac{x}{1+x}$ on $[1, 3]$

20. $(8 - x)^{1/3}$

21. $x(x-1)(x-2), x \in [0, 1/2]$

22. Prove that, $|\ln x - \ln y| \le 3|x - y|$ for $\frac{1}{3} \le y < x \le 3$, by using Mean Value Theorem.

23. Determine value of c for $f(x) = x^3$ and $g(x) = \tan^{-1}x$ on interval $[0, 1]$ which satisfy the Cauchy Mean Value Theorem.

24. Find value of c for $f(x) = 1/x$ and $g(x) = 1/x^2$ on interval $[4, 6]$ which satisfy the Cauchy Mean Value Theorem.

25. Determine value of c for $f(x) = |x - 1|$ and $g(x) = \ln x$ on interval $[2, 5]$ which satisfy the Cauchy Mean Value Theorem.

26. Prove that $1 - \dfrac{x^2}{2} < \cos x$, for $x \neq 0$ using CMVT.

Evaluate the following limits:

27. $\lim\limits_{x \to 0} \dfrac{\sin x - x}{x^2}$

28. $\lim\limits_{x \to 0} \dfrac{\ln(\sec x)}{3x^2}$

29. $\lim\limits_{x \to \infty} \dfrac{5x - 8}{4x + 7}$

30. $\lim\limits_{x \to 0^+} \dfrac{\ln(e^x - 1)}{\ln(x)}$

31. $\lim\limits_{x \to 0^+} x \ln x$

32. $\lim\limits_{x \to 0^+} x^x$

33. $\lim\limits_{x \to 0^+} \ln x^{1/x}$

34. $\lim\limits_{x \to 1^+} x^{1/(x-1)}$

35. $\lim\limits_{x \to 0^+} \left(\dfrac{1}{\sin x} - \dfrac{1}{x}\right)$

Answers:

1. 2, –4
2. $\pm\frac{1}{\sqrt{2}}$
3. $5\pi/4$
4. –2
5. 1
6. –1, –2
7. $\pi/4 + 2k\pi$
8. No relative extrema
9. $x = 0$ is relative max and $x = 1$ is relative min
10. $x = 15$ is absolute max and $x = -6.9016$ is absolute min
11. $x = 1$ is absolute max and $x = -2$ is absolute min
12. $x = 2$ is absolute max and $x = -2$ is absolute min
13. $c = \pm\frac{1}{\sqrt{3}}$
14. $c = -4$
15. $c = \sqrt{ab}$
16. $c = \frac{3\pi}{4}$
17. $b = 2$
18. $c = 1$
19. 1.8284
20. $8 - \frac{8}{3\sqrt{3}}$
21. 0.2362 or $1 - \sqrt{21}/6$
22. Prove
23. $c = 0.60$
24. $c = 1.8$
25. $c = 3.27$
26. Hint: $f(x) = 1 - \cos x$ and $x^2/2$
27. 0
28. 1/6
29. 5/4
30. 1
31. 0
32. 1
33. e
34. e
35. 0

4 Sequence and Series

In this chapter, we will talk about different arithmetic topics. A sequence is the set of ordered elements leading to some rule while a series is the sum of sequential elements. One can also say sequence and series include the formula of nth term and sum.

4.1 SEQUENCES

A sequence is an arrangement of numbers in a definite order followed by some rule. This list of arrangements may contain infinite terms. Terms are denoted by $a_1, a_2, a_3, ..., a_n, a_{n+1}, ...$, where subscripts denote the position of the term, i.e., first term, second term, ..., n^{th} term, $(n + 1)^{st}$ term and so on.

We should be very clear about this subscript that $(n + 1)^{st}$ term is the next term after the nth term not like adding 1 in n^{th} term i.e., $a_{n+1} \neq a_n + 1$.

The sequence is denoted by $\{a_1, a_2, ..., a_n, a_{n+1}, ...\}$ or $\{a_n\}$ or $\{a_n\}_{n=1}^{\infty}$ or $\langle a_n \rangle$ or (a_n).

Here, a_n is nothing but a generalized formula of sequence. Thus, a sequence can be written in terms of function as $f: N \to X$ defined by $f(n) = x_n$, $\forall n \in \mathrm{N}$. Here, if the range is a subset of $\mathbb{R}$ (real numbers) then it is considered as a **Real Sequence**.

4.1.1 Arithmetic Progression (A.P.)

It is a sequence in which except initial term each term is obtained by adding a fixed number to the previous term. If a sequence a_n is arithmetic then there is a fixed number d so that $a_{n+1} - a_n = d$ for any $n \in \mathrm{N}$. The number d is the **difference** and a_1 is the first term. Now, we have n^{th} term as $a_n = a_{n-1} + d$.

Again, we have, $a_{n-1} = a_{n-2} + d \Rightarrow a_n = a_{n-2} + d + d = a_{n-2} + 2d$.

In this manner, we get:

$$a_n = a_{n-1} + d = a_{n-2} + 2d$$

$$\Rightarrow a_n = a_{n-3} + d + 2d = a_{n-3} + 3d$$

$$\vdots$$

$$\Rightarrow a_n = a_2 + (n-2)d$$

$$\Rightarrow a_n = a_1 + (n-1)d \text{ which is the formula for } a_n.$$

The sum S_n of the first n-terms is given by, $S_n = \frac{n}{2}[2a + (n-1)d]$.

If a_1, a_2, a_3, ... are in arithmetic progression having differences then,

(i) $a_1 \pm c, a_2 \pm c, a_3 \pm c, \ldots$ are also in A.P. with a difference d.

(ii) $a_1c, a_2c, a_3c, \ldots$ are also in A.P. with a difference $cd (c \neq 0)$.

(iii) $\frac{a_1}{c}, \frac{a_2}{c}, \frac{a_3}{c}, \ldots$ are also in A.P. with difference $\frac{d}{c}(c \neq 0)$.

If $a_1, a_2, a_3, \ldots$ and $b_1, b_2, b_3, \ldots$ are two in arithmetic progression (A.P.)., then

(i) $a_1 \pm b_1, a_2 \pm b_2, a_3 \pm b_3, \ldots$ are also in A.P.

(ii) $a_1b_1, a_2b_2, a_3b_3, \ldots$ and $\frac{a_1}{b_1}, \frac{a_2}{b_2}, \frac{a_3}{b_3}, \ldots$ are not in A.P.

If a_1, a_2, a_3, are two in arithmetic progression (A.P.) then the arithmetic mean is given by, $a_2 = \frac{a_1 + a_3}{2}$.

4.1.2 Geometric Progression (G.P.)

It is a sequence in which except initial term each term is obtained by multiplying the previous term with a non-zero constant. If a sequence a_n is geometric then there is a fixed number r so that $\frac{a_n}{a_{n-1}} = r$ for any $n \in \mathbb{N}$. The number r is a common ratio of consecutive terms. Now, we have n^{th} term as $a_n = ra_{n-1}$ and a_1 is the first term.

$$a_n = ra_{n-1} = r(ra_{n-2}) = r^2 a_{n-2}$$

$$\Rightarrow a_n = r(r^2 a_{n-2}) = r^3 a_{n-3}$$

$$\vdots$$

$$a_n = ra_{n-1} = r(ra_{n-2}) = r^2 a_{n-2}$$

Thus, terms will have $a, ar, ar^2, \ldots, ar^{n-1}$ having common ratio $r = \frac{ar^{n-1}}{ar^{n-2}}$ and n^{th} term as $a_n = ar^{n-1}$.

The sum S_n of the first n-terms is given by, $S_n = \dfrac{a(r^n - 1)}{r-1}$, if $r \neq 1$.

If a_1, a_2, a_3 are two in geometric progression (A.P.) then the geometric mean is given by, $a_2 = \sqrt{a_1 a_3}$.

Results:

(i) If $a_1, a_2, a_3, \ldots$ are in A.P., then for $x > 0$, $x^{a_1}, x^{a_2}, x^{a_3}, \ldots$ are in G.P.

(ii) If $a_1, a_2, a_3, \ldots$ and $b_1, b_2, b_3, \ldots$ are two in G.P., then $a_1b_1, a_2b_2, a_3b_3, \ldots$ and $\dfrac{a_1}{b_1}, \dfrac{a_2}{b_2}, \dfrac{a_3}{b_3}, \ldots$ are also in G.P.

Now, let us take a simple example of sequence $s\{a_n\} = \left\{\dfrac{n+1}{n^2}\right\}_{n=1}^{\infty}$. Then sequential terms are $2, \dfrac{3}{4}, \dfrac{4}{9}, \dfrac{5}{16}, \dfrac{6}{25}, \dfrac{7}{36}, \ldots$. Let's understand this sequence by graph (Fig. 4.1). One can see that, as the value of n increases sequence tends to zero. i.e., $\lim\limits_{n\to\infty} a_n = \lim\limits_{n\to\infty} \left\{\dfrac{n+1}{n^2}\right\} = 0$.

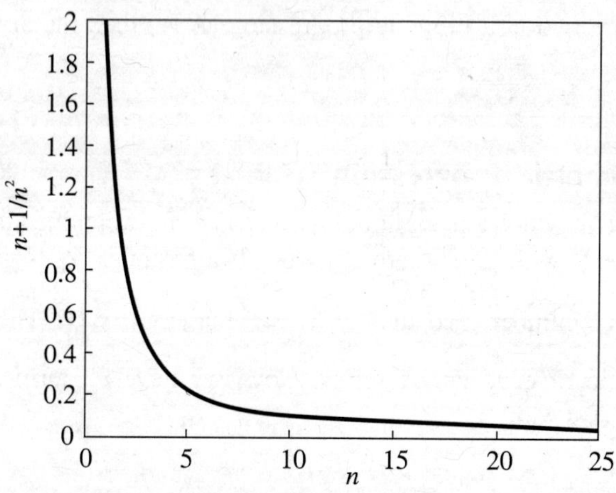

FIGURE 4.1 Graph of $\left\{\dfrac{n+1}{n^2}\right\}_{n=1}^{\infty}$.

For any sequence a_n if we have $\lim\limits_{n\to\infty} a_n = L$, we say that as n approaches to infinity, a_n approaches to L.

If we have $\lim\limits_{n\to\infty} a_n = \infty$ or $\lim\limits_{n\to\infty} a_n = -\infty$, we say that as n approaches to infinity, a_n, the values of a_n get positive or negative respectively larger and larger value in without bound.

If the limit exists and has a finite value, all the sequence terms must go to closer and closer to that finite value as n increases. In mathematical terminology, we say that if the limit $\lim_{n\to\infty} a_n$ exists and is finite then the sequence **converges** to that finite limit. If this limit doesn't exist or is infinite then the sequence **diverges**. We will use the term the sequence is **convergent** or **divergent**. A sequence that neither converges nor diverges is called an **oscillatory** sequence.

Definition: Suppose that $\{a_n\}_{n=1}^{\infty}$ is a sequence of real numbers. We say that $\lim_{n\to\infty} a_n = L$ if for every $\in\, > 0$ there is an $N > 0$ so that, whenever $n \geq N$, $|a_n - L| < \in$. If $\lim_{n\to\infty} a_n = L$ we say that the sequence converges, otherwise it diverges.

Thus, we can compare function $f(x)$ as a sequence, $f(n) = a_n$ and $\lim_{x\to\infty} f(x) = L$ implies $\lim_{n\to\infty} a_n = L$.

EXAMPLE 4.1 Determine whether the sequence is convergent or divergent.

(i) $\left\{\dfrac{n^2 - 3n + 5}{7n^3 + 4}\right\}$

Ans: (i) $$\lim_{n\to\infty} \frac{n^2 - 3n + 5}{7n^3 + 4} = \frac{\lim_{n\to\infty} \dfrac{n^2 - 3n + 5}{n^3}}{\lim_{n\to\infty} \dfrac{7n^3 + 4}{n^3}} = \frac{\lim_{n\to\infty}\left(\dfrac{1}{n} - \dfrac{3}{n^2} + \dfrac{5}{n^3}\right)}{\lim_{n\to\infty}\left(7 + \dfrac{4}{n^3}\right)} = \frac{0 - 0 + 0}{7 + 0} = 0$$

The sequence converges to 0.

(ii) $5 + (-1)^n \dfrac{1}{n}$

Ans: $$\lim_{n\to\infty}\left(5 + (-1)^n \frac{1}{n}\right) = 5 + 0 = 5$$

The sequence converges to 5.

As we have already studied the limit properties. If two sequences $\{a_n\}$ and $\{b_n\}$ are convergent then,

(a) $\lim_{n\to\infty} (a_n \pm b_n) = \lim_{n\to\infty} a_n \pm \lim_{n\to\infty} b_n$

(b) $\lim_{n\to\infty} (a_n b_n) = (\lim_{n\to\infty} a_n)(\lim_{n\to\infty} b_n)$

(c) $\lim_{n\to\infty}\left(\dfrac{a_n}{b_n}\right) = \dfrac{\lim_{n\to\infty} a_n}{\lim_{n\to\infty} b_n}$, provided $\lim_{n\to\infty} b_n \neq 0$.

(d) $\lim_{n\to\infty} ca_n = c \lim_{n\to\infty} a_n$

(e) $\lim_{n\to\infty} a_n^p = [\lim_{n\to\infty} a_n]^p$, provided $a_n \geq 0$.

Some results:

(i) If $\lim_{n\to\infty} |a_n| = 0$ then $\lim_{n\to\infty} a_n = 0$.

(ii) **Sandwich/Squeeze rule for the sequence:** If $a_n \leq c_n \leq b_n$ for all $n > N$ and $\lim_{n\to\infty} a_n = \lim_{n\to\infty} b_n = L$ then $\lim_{n\to\infty} c_n = L$.

(iii) **Convergence of Geometric sequence:** The sequence $\{a^n\}_{n=0}^{\infty}$ converges if $-1 < a \leq 1$ and diverges for $a \geq 1$. Also,
$$\lim_{n\to\infty} a^n = \begin{cases} 0 & -1 < a < 1 \\ 1 & a = 1 \end{cases}.$$

(iv) **Convergence of some sequences:**

(a) $\lim_{n\to\infty} \sqrt[n]{n} = 1$

(b) $\lim_{n\to\infty} x^{1/n} = 1$

(c) $\lim_{n\to\infty} x^n = 0$

(d) $\lim_{n\to\infty} \left(1 + \frac{x}{n}\right)^n = e^x$

EXAMPLE 4.2 Check whether $\left\{\frac{\ln n}{n}\right\}_{n=1}^{\infty}$ converges or diverges.

Ans: $\lim_{n\to\infty} \frac{\ln n}{n} = \lim_{n\to\infty} \frac{1/n}{1} = 0$ (using L'Hospital Rule)

EXAMPLE 4.3 Determine whether $\{(-1)^n\}_{n=0}^{\infty}$ converges or diverges.

Ans: The sequence is 1, –1, 1 – 1, 1, ...

This sequence does not make sense, it oscillates finitely.

EXAMPLE 4.4 The sequence $\{(-1)^n n\}_{n=1}^{\infty}$.

Ans: The sequence is $\{(-1)^n n\}_{n=1}^{\infty} = -1, 2, -3, 4, \ldots$ unbounded and oscillates infinitely.

- One more thing is that some sequences are increasing in nature while others are decreasing in nature. Thus, for any sequence $\{a_i\}_{n=1}^{\infty}$ if $a_i \leq a_{i+1}$ for all i then it is **increasing** and if $a_i < a_{i+1}$ for all i then it is **strictly increasing** sequence. While for any sequence $\{a_i\}_{n=1}^{\infty}$ if

$a_i \geq a_{i+1}$ for all i then it is **decreasing** and if $a_i > a_{i+1}$ for all i then it is **strictly decreasing** sequence. If the sequence has any of these characteristics, then it is called **monotonic**.

- A sequence $\{a_n\}$ is said to be **bounded above** if $\exists M \in \mathbb{R}$ such that $a_n \leq M$ for $\forall n$. It is **bounded below** if $\exists m \in \mathbb{R}$ such that $a_n \geq m$ for $\forall n$. And it is bounded if it is bounded both below and above.
- For example, $\{n\}$ is strictly increasing and bounded below by 1. $\{1/n\}$ is strictly decreasing and bounded below by 0 and bounded above by 1. $\{(-1)^n/n\}_{n=1}^{\infty}$ is neither decreasing nor increasing.

4.2 SUBSEQUENCE

A subsequence of a given sequence is a sequence that can be obtained from the given sequence by removing some or no terms without altering the order of terms. Let $\{a_n\}_{n\geq 1}$ be a sequence and $\{n_k\}_{k\geq 1}$ any increasing sequence then the sequence $\{a_{n_k}\}_{k\geq 1}$ is called subsequence of $\{a_n\}_{n\geq 1}$. For example, if we consider $\left\{\frac{1}{n}\right\}$ is sequence then $\left\{\frac{1}{n^2}\right\}$, $\left\{\frac{1}{2n}\right\}$, $\left\{\frac{1}{3n+1}\right\}$, $\cdots$ are some subsequence. A few facts about subsequence are:

- Every sequence is a subsequence of itself.
- If $a_n \to L$ (converges to some limit), then every subsequence $a_{n_k} \to L$ of it (converges to the same limit). If $a_n \to \infty$ then every subsequence $a_{n_k} \to \infty$ as well.
- The sequence $\{a_n\}$ is divergent if it has two convergent sequences with different limits. Also, it is divergent if it's a subsequence tends to ∞ or subsequence tends to $-\infty$.
- A sequence is convergent iff $\exists L \in \mathbb{R}$ such that every subsequence of the given sequence has a further subsequence that converges to L.

THEOREM 4.1 (BMCT: Bounded Monotone Convergence Theorem) Every bounded monotonic sequence of real numbers converges.

Proof: Let $\{a_n\}$ be an increasing sequence that is bounded above.

$\therefore$ $a_n \leq M$ for all $n \in \mathbb{N}$ and for given $\varepsilon > 0$ $\exists N$ such that $a_N > a - \varepsilon$.

Since, $a - \varepsilon < a_N \leq a_n \leq a \leq a + \varepsilon$ for $n \geq \mathbb{N}$.

That is, $|a_n - a| < \varepsilon$ for $n \geq \mathbb{N}$.

Thus, by definition we can say that, $\{a_n\}$ converges to its least upper bound. $\lim\limits_{n\to\infty} a_n = a$.

Identically, the decreasing sequence converges to its greatest upper bound.

THEOREM 4.2 Every monotonic sequence that is not bounded must diverge to ∞ or −∞.

Proof: Let $\{a_n\}$ be an increasing sequence that is not bounded above.

As its first term is a_1, it is bounded below by a_1.

Since $\{a_n\}$ is an increasing sequence, given $M > 0$, $\exists N$ such that $a_n \geq a_N > M$ for $n \geq N$.

Thus, it follows that $\lim_{n \to \infty} a_n = \infty$.

Identically proof for decreasing sequence.

This completes the proof.

LEMMA 4.1 Every sequence of real numbers contains a monotonic subsequence.

Proof: Let $\{a_n\}$ be a sequence of real numbers. If the sequence has a monotone subsequence, then $\{a_n\}$ has either infinitely many or only finitely many peak points.

Assume that there are n infinitely many peak points. Let n_1, n_2, n_3, ... be the peak points at where n_1 be the smallest peak point and $n_2 > n_1$.

(i) $a_{n_1} \geq a_k$ for all $k \in \mathbb{N}$ with $k \geq n_1$

(ii) $a_{n_2} \geq a_k$ for all $k \in \mathbb{N}$ with $k \geq n_2$

From these two we say that, $a_{n_1} \geq a_{n_2}$.

Continuing in this way, if we have n_{k+1} and n_k points then from mathematical induction we have $a_{n_k} \geq a_{n_{k+1}}$ implies a monotonically decreasing subsequence of sequence a_n.

Similarly, if the sequence has finitely many n points, then $a_n \geq a_k$ for all $k \in \mathbb{N}$ with $k \geq n$.

Now, we choose another subsequence having, t_1 greater than all peak points such that t_1 itself is not a peak point. There exists t_2 which is also not a peak point for which $a_{t_1} < a_{t_2}$.

Continuing this way, if we have t_{k+1} and t_k points then from mathematical induction we have $a_{t_k} < a_{t_{k+1}}$ implies a_{t_k} is a monotonically increasing subsequence of sequence a_n.

Thus, it completes the proof.

THEOREM 4.3 (Bolzano-Weierstrass Theorem): Every bounded sequence of real numbers has a convergent subsequence.

Proof: Let $\{a_n\}$ be a bounded sequence of real numbers.

Hence, from lemma 4.1 we can say that, it has a monotonic subsequence say $\{a_{n_k}\}$.

As $\{a_n\}$ is bounded, so as $\{a_{n_k}\}$ and by BMCT $\{a_{n_k}\}$ converges.

This completes the proof.

EXAMPLE 4.5 Check convergence of the sequence $a_1 = 1$, $a_{n+1} = \sqrt{2+a_n}$ for $n \geq 1$.

Ans: Here, $a_n > 0$. Since $a_1 < 2$ we can say that, $a_{n+1} = \sqrt{2+a_n} < \sqrt{2+2} = 2$ for all $n \geq 1$.

Now, $a_{n+1} - a_n = \sqrt{2+a_n} - a_n \geq 0$ implies $(2-a_n)(1+a_n) \geq 0$.

Thus, $a_n \leq 2$ tells that given sequence is bounded above by 2 and monotonically increasing implies its convergence.

$$\lim_{n\to\infty} a_{n+1} = \lim_{n\to\infty} \sqrt{2+a_n} = \sqrt{2+a} \text{ which gives } (a-2)(a+1) = 0 \Rightarrow a = 2.$$

EXAMPLE 4.6 Check the convergence of the sequence $\left\{\frac{n!}{n^n}\right\}$.

Ans: Here, $\frac{a_{n+1}}{a_n} = \frac{(n+1)!}{(n+1)^{n+1}} \times \frac{n^n}{n!} = \frac{(n+1)!}{n!} \times \frac{n^n}{(n+1)^{n+1}}$

$$\Rightarrow \frac{a_{n+1}}{a_n} = \frac{(n+1)n!}{n!} \times \frac{n^n}{(n+1)^n(n+1)}$$

$$\Rightarrow \frac{a_{n+1}}{a_n} = \frac{n^n}{(n+1)^n} < 1.$$

Thus, $a_{n+1} < a_n$ implies sequence is decreasing and it is bounded hence it converges.

4.3 CAUCHY SEQUENCE

A sequence $\{a_n\}_{n=0}^{\infty}$ is called a Cauchy sequence if for given $\varepsilon > 0$, $\exists m \in \mathbb{N}$ such that $|a_n - a_m| < \varepsilon$ for every $n \geq m$.

Equivalently, we say that sequence $\{a_n\}_{n=0}^{\infty}$ is Cauchy if for each $\varepsilon > 0$, $\exists m \in \mathbb{N}$ such that $|a_{n+p} - a_n| < \varepsilon$ for every $n \geq m$ and $p \in \mathbb{N}$.

THEOREM 4.4 Completeness Criterion for Sequences
A sequence is convergent if it is a Cauchy sequence.

Proof: Let $\{a_n\}_{n=0}^{\infty}$ be a convergent sequence. Therefore, for given $\varepsilon > 0$ there exists m_0 such that, $|a_n - a_m| = |a_n - l + l - a_m| \leq |a_n - l| + |l - a_m| < \frac{\varepsilon}{2} + \frac{\varepsilon}{2} < \varepsilon$, $\forall n, m \geq m_0$.

Implies $\{a_n\}_{n=0}^{\infty}$ is Cauchy sequence.

Conversely, let $\{a_n\}_{n=0}^{\infty}$ is a Cauchy sequence in $\mathbb{R}$.

Therefore, from definition $\{a_n\}_{n=0}^{\infty}$ is a convergent sequence.

Hence the proof.

THEOREM 4.5 Cauchy sequences are bounded.

Proof: By Theorem 4.4 every Cauchy sequence is convergent and convergent sequence is bounded.

Hence the proof.

EXAMPLE 4.7 Shows the convergence of $\{a_n\}$ where $a_n = \sum_{i=0}^{n} \frac{1}{n}$.

Ans: Suppose that the given is convergent implying the sequence is Cauchy. Hence according to the definition, for given $\varepsilon > 0$ there exists m such that,

$$n \geq m \Rightarrow \left|S_{n+p} - S_n\right| < \varepsilon, p \in \mathbb{N}$$

Let's take $\varepsilon = \frac{1}{2}$

$$\left|S_{n+p} - S_n\right| < \frac{1}{2}$$

$$\therefore \left|\left(1 + \frac{1}{2} + \frac{1}{3} + \ldots + \frac{1}{n} + \frac{1}{n+1} + \ldots + \frac{1}{n+p}\right) - \left(1 + \frac{1}{2} + \frac{1}{3} + \ldots + \frac{1}{n}\right)\right| < \frac{1}{2}$$

$$\therefore \left|\frac{1}{n+1} + \ldots + \frac{1}{n+p}\right| < \frac{1}{2}$$

Separately if we take $p = n$ then,

$$\left|\frac{1}{n+1} + \frac{1}{n+2} + \ldots + \frac{1}{n+n}\right| > \left|\frac{1}{n+n} + \frac{1}{n+n} + \ldots + \frac{1}{n+n}\right| \quad (n \text{ times})$$

$$= \left|\frac{1}{2n} + \frac{1}{2n} + \ldots + \frac{1}{2n}\right| = \left|n\frac{1}{2n}\right| = \left|\frac{1}{2}\right|$$

$\left|S_{n+p} - S_n\right| > \frac{1}{2}$ which contradicts.

Hence, our assumption is wrong. Given sequence is not convergent.

4.4 SERIES

A Series is the sum of the terms in a defined sequence. If $\{a_n\}_{n=0}^{\infty}$ is a sequence then the associated series is $\sum_{i=0}^{\infty} a_n = a_0 + a_1 + a_2 + a_3 + \ldots$. Associated with a series is a second sequence, called the **sequence of partial sums** $\lim_{n \to \infty} s_n$ $\{s_n\}_{n=0}^{\infty}$, $s_n = \sum_{i=0}^{n} a_i$ or simply $\sum a_n$.

So, $s_0 = a_0$, $s_1 = a_0 + a_1$, $s_2 = a_0 + a_1 + a_2$, $s_3 = a_0 + a_1 + a_2 + a_3$, ...,

$s_n = a_0 + a_1 + \ldots + a_n$

4.5 CONVERGENCE AND DIVERGENCE OF SERIES

Let $\sum a_n$ be an infinite series. As $n \to \infty$, there are three possibilities:

(i) **Convergent Series:** $\sum a_n$ is convergent to limit l, if the sequence $\{a_n\}$ is convergent, i.e. $\lim_{n \to \infty} a_n = l$, this limit exists and is finite or unique.

(ii) **Divergent Series:** $\sum a_n$ is divergent, if the sequence $\{a_n\}$ is divergent, i.e., $\lim_{n \to \infty} a_n = \pm\infty$, this limit does not exist or infinity.

(iii) **Oscillatory Series:** There are two situations. The series oscillates finitely or infinitely.

- $\sum a_n$ oscillates finitely, if the sequence $\{a_n\}$ oscillates finitely. i.e., $\lim_{n \to \infty} a_n = l$, this limit is finite but not unique.
- $\sum a_n$ oscillates infinitely, if the sequence $\{a_n\}$ oscillates infinitely. i.e., $\lim_{n \to \infty} a_n = \pm\infty$.

Note: Divergent or Oscillate Series is often called a non-convergent series.

Properties:

If $\sum a_n$ and $\sum b_n$ are both convergent series then,

- $\sum_{n=k}^{\infty} a_n \pm \sum_{n=k}^{\infty} b_n = \sum_{n=k}^{\infty} (a_n \pm b_n)$ is also convergent.
- $\sum ca_n$ is also convergent.

EXAMPLE 4.8 Determine the value of the series if it converges, $\sum_{n=0}^{\infty} (-1)^n$.

Ans: Here, $\sum_{n=0}^{\infty} (-1)^n$.

Here, $s_0 = 1$, $s_1 = 1 - 1 = 0$, $s_2 = 1 - 1 + 1 = 1$, etc.

$\{s_n\} = \{1, 0, 1, 0, 1, \ldots\}$ will give $\lim_{n\to\infty} s_n$ doesn't exist.

Hence, the series diverges.

4.6 SPECIAL SERIES

4.6.1 Geometric Series

A Geometric series can be written as $\sum_{n=0}^{\infty} ar^n$.

The partial sum for this series is $s_n = \dfrac{a(1-r^n)}{1-r} = \dfrac{a}{1-r} - \dfrac{ar^n}{1-r}$.

(i) For convergence of this series, let's take the limit of the partial sum.

$$\lim_{n\to\infty} s_n = \lim_{n\to\infty}\left(\frac{a}{1-r} - \frac{ar^n}{1-r}\right) = \lim_{n\to\infty}\frac{a}{1-r} - \lim_{n\to\infty}\frac{ar^n}{1-r} = \frac{a}{1-r} - \frac{a}{1-r}\lim_{n\to\infty}(r^n)$$

This limit exists, if $-1 < r < 1$ implies $|r| < 1$ then $\lim_{n\to\infty}(r^n)$ will converge to zero.

$$\lim_{n\to\infty} s_n = \frac{a}{1-r}$$

(ii) Series is divergent if $r \geq 1$.

$s_n = \dfrac{r^n - 1}{r-1}$, here $s_n \to \infty$ as $n \to \infty$ implies series is divergent.

Finitely oscillating if $r = -1$ and infinitely oscillating if $r < -1$.

EXAMPLE 4.9 Show the convergence of $1 - \dfrac{1}{2} + \dfrac{1}{2^2} - \dfrac{1}{2^3} + \dfrac{1}{2^4} - \ldots$.

Ans: Since from the given series the value of $|r| = \left|-\dfrac{1}{2}\right| = \dfrac{1}{2} < 1$, the series is convergent.

EXAMPLE 4.10 Determine if the series $\sum_{n=0}^{\infty}\left(\frac{2^n-1}{3^n}\right)$ is convergent or divergent, and if it is convergent then find its sum.

Ans: Here,

$$\sum_{n=0}^{\infty}\left(\frac{2^n-1}{3^n}\right)=\sum_{n=0}^{\infty}\left(\frac{2^n}{3^n}-\frac{1}{3^n}\right)=\sum_{n=0}^{\infty}\left(\frac{2^n}{3^n}\right)-\sum_{n=0}^{\infty}\left(\frac{1}{3^n}\right)=\sum_{n=0}^{\infty}\left(\frac{2}{3}\right)^n-\sum_{n=0}^{\infty}\left(\frac{1}{3}\right)^n$$

Geometric series will converge $\lim_{n\to\infty} s_n=\frac{a}{1-r}$, so the given series will converge to

$$\sum_{n=0}^{\infty}\left(\frac{2}{3}\right)^n=\frac{1}{1-\frac{2}{3}}=3 \text{ and } \sum_{n=0}^{\infty}\left(\frac{1}{3}\right)^n=\frac{1}{1-\frac{1}{3}}=\frac{3}{2}$$

Therefore, $\sum_{n=0}^{\infty}\left(\frac{2^n-1}{3^n}\right)=\sum_{n=0}^{\infty}\left(\frac{2}{3}\right)^n-\sum_{n=0}^{\infty}\left(\frac{1}{3}\right)^n=3-\frac{3}{2}=\frac{3}{2}$.

4.6.2 Harmonic Series

It is an infinite series. $\sum_{n=1}^{\infty}\frac{1}{n}$ is formed by summing all unit fractions. This series is known as the harmonic series.

Here, $\lim_{n\to\infty}\frac{1}{n}=0$ implies the convergence of series but here scenario will differ.

Let's find $\sum_{n=1}^{10}\frac{1}{n}\approx 2.93$, $\sum_{n=1}^{1000}\frac{1}{n}\approx 7.49$ seems that, the partial sum becomes larger and tends to infinity. If we take,

$$1+\frac{1}{2}+\frac{1}{3}+\frac{1}{4}>1+\frac{1}{2}+\frac{1}{4}+\frac{1}{4}>1+\frac{1}{2}+\frac{1}{2}$$

$$1+\frac{1}{2}+\frac{1}{3}+\frac{1}{4}+\frac{1}{5}+\frac{1}{6}+\frac{1}{7}+\frac{1}{8}>1+\frac{1}{2}+\frac{1}{4}+\frac{1}{4}+\frac{1}{8}+\frac{1}{8}+\frac{1}{8}+\frac{1}{8}>1+\frac{1}{2}+\frac{1}{2}+\frac{1}{2}$$

$$1+\frac{1}{2}+\frac{1}{3}+\ldots+\frac{1}{16}>1+\frac{1}{2}+\frac{1}{4}+\frac{1}{4}+\frac{1}{8}+\ldots+\frac{1}{8}+\frac{1}{16}+\ldots+\frac{1}{16}>1+\frac{1}{2}+\frac{1}{2}+\frac{1}{2}+\frac{1}{2}$$

and so on.

Similar manner, $1+\frac{1}{2}+\frac{1}{3}+\ldots+\frac{1}{2^n}>1+\frac{n}{2}$

Thus, the sum will be as big as we like neither finite nor unique implies that the series is divergent.

4.6.3 Alternating Series

Alternating series will include both positive and negative terms in an alternate sense. Let us consider the series $\{a_n\}=\left\{\frac{(-1)^{n-1}}{n}\right\}$ having partial sums s_1, s_2, s_3, s_4, s_5, ... and so on.

$$\sum_{n=1}^{\infty}\frac{(-1)^{n-1}}{n}=\frac{1}{1}+\frac{-1}{2}+\frac{1}{3}+\frac{-1}{4}+\frac{1}{5}\ldots=\frac{1}{1}-\frac{1}{2}+\frac{1}{3}-\frac{1}{4}+\frac{1}{5}\ldots$$

Here, $|a_n|$ forms a decreasing sequence. In this alternating series, partial sums s_1, s_3, s_5, ... and so on form one decreasing sequence bounded below by s_2 likewise s_2, s_4, s_6, ... and so on forms another decreasing sequence bounded above by s_1 shows the convergence of these two sequences. As the value of a_i is becoming smaller, these two sequences converge to the same value hence the entire sequence converges too.

- Suppose we have a series with $a_n = (-1)^n b_n$, $b_n \geq 0$ for all n. Then if, $\{b_n\}$ is decreasing sequence and $\lim_{n\to\infty} b_n = 0$ then $\sum a_n$ is convergent.

4.7 TEST ON SERIES

4.7.1 Integral Test

Suppose that $f(x)$ is a continuous, positive, and decreasing function on the interval $[k, \infty)$ and that $f(n) = a_n$ then,

1. If $\int_k^{\infty} f(x)\,dx$ is convergent so is $\sum_{n=k}^{\infty} a_n$.

2. If $\int_k^{\infty} f(x)\,dx$ is divergent so is $\sum_{n=k}^{\infty} a_n$.

EXAMPLE 4.11 Check the convergence of $\sum_{n=1}^{\infty}\frac{n}{n^2+1}$.

Ans: The sequence $\{a_n\}=\left\{\frac{n}{n^2+1}\right\}$ is positive and decreasing. If we consider $f(x)=\frac{x}{x^2+1}$ then it is a continuous function.

$$\int_1^{\infty} \frac{x}{x^2+1} dx = \lim_{k\to\infty} \int_1^{t} \frac{x}{x^2+1} dx$$

$$\therefore \quad \int_1^{\infty} \frac{x}{x^2+1} dx = \lim_{t\to\infty} \frac{1}{2} \ln\left|x^2+1\right|_1^t = \lim_{t\to\infty}\left[\frac{1}{2}\ln\left|t^2+1\right| - \frac{1}{2}\ln|2|\right] = \infty$$

Hence, the series diverges.

EXAMPLE 4.12 Check the convergence of $\sum_{n=0}^{\infty} ne^{-n^2}$.

Ans: The function $f(x) = xe^{-x^2}$ is positive. For the nature of the function, we have,

$$f'(x) = e^{-x^2}(1 - 2x^2)$$

$$\therefore \quad f'(x) = 0 = e^{-x^2}(1 - 2x^2)$$

Then, the function is increasing on $\left[0, \frac{1}{\sqrt{2}}\right]$ and decreasing on $\left[\frac{1}{\sqrt{2}}, \infty\right)$, so ultimately the function will decrease.

$$\int_0^{\infty} f(x)\,dx = \lim_{k\to\infty} \int_0^{k} xe^{-x^2} dx = \lim_{k\to\infty}\left[-\frac{1}{2}e^{-x^2}\right]_0^t = \lim_{k\to\infty}\left[\frac{1}{2} - \frac{1}{2}e^{-k^2}\right] = \frac{1}{2}$$

Hence, the series is convergent.

4.7.2 *p*-series Test

The infinite series, $\sum_{n=1}^{\infty} \frac{1}{n^p} = \frac{1}{1^p} + \frac{1}{2^p} + \frac{1}{3^p} + \dots$ is

(i) Convergent when $p > 1$.

(ii) Divergent when $p \le 1$.

EXAMPLE 4.13 Check the convergence of $\sum_{n=1}^{\infty} \frac{7}{n^3}$ and $\sum_{n=1}^{\infty} \frac{3}{\sqrt{n}}$.

Ans: From p-series Test one can easily observe that, $\sum_{n=1}^{\infty} \frac{7}{n^3}$ is p-series with $p > 1$ implies it is convergent while $\sum_{n=1}^{\infty} \frac{3}{\sqrt{n}}$ is p-series with $p < 1$ implies it is divergent.

4.7.3 Comparison Test and Limit Comparison Test

(a) *Comparison Test*

If $\sum a_n$ and $\sum b_n$ be two series with a_n, $b_n \geq 0$ for all n and $a_n \leq b_n$ for all n. Then,

1. If $\sum b_n$ is convergent then so is $\sum a_n$.
2. If $\sum a_n$ is divergent then so is $\sum b_n$.

EXAMPLE 4.14 Check the convergence of $\sum_{n=1}^{\infty} \frac{n^3+3}{n^6+7}$.

Ans: If we say $b_n = \frac{n^3+3}{n^6+7}$, we can find another series for applying this comparison test.

So, either one can decrease the power in the denominator or increase the power in the numerator.

$$\frac{n^3+3}{n^6+7} < \frac{n^3+3}{n^6}$$

Further, $\sum_{n=1}^{\infty} \frac{n^3+3}{n^6} = \sum_{n=1}^{\infty} \frac{n^3}{n^6} + \sum_{n=1}^{\infty} \frac{3}{n^6} = \sum_{n=1}^{\infty} \frac{1}{n^3} + \sum_{n=1}^{\infty} \frac{3}{n^6}$

As, these both series convergent p-series Test, sum of two series also convergent.

Hence, by comparison test given series is also convergent.

EXAMPLE 4.15 Check the series $\sum_{n=1}^{\infty} \frac{e^{-n}}{n}$.

Ans: We can write for this series as $\frac{e^{-n}}{n} < \frac{e^{-n}}{1} = e^{-n}$.

Now, we have to check the convergence of $\sum_{n=1}^{\infty} e^{-n}$.

$$\int_1^{\infty} e^{-x} dx = \left[-e^{-x}\right]_1^{\infty} = e^{-1}$$

As this integral is convergent, the given series is also convergent.

(b) *Limit Comparison Test*

Suppose $\sum a_n$ and $\sum b_n$ be two series with $a_n \geq 0, b_n > 0$ for all n, then we define the term $L = \lim_{n\to\infty} \frac{a_n}{b_n}$. If $L > 0$ and finite then either both series convergent or both divergent.

EXAMPLE 4.16 Check the series $\sum_{n=0}^{\infty} \frac{3}{5^n - n}$.

Ans: For this limit comparison test, we have to choose or find a series that can be helpful to check the convergence of $\sum a_n$.

Let $\sum b_n = \sum_{n=0}^{\infty} \frac{1}{5^n}$.

Then,
$$L = \lim_{n\to\infty} \frac{a_n}{b_n} = \lim_{n\to\infty} \frac{3/(5^n - n)}{3/5^n}$$

$\therefore$
$$L = \lim_{n\to\infty} \frac{5^n - n}{5^n} = \lim_{n\to\infty} \left(1 - \frac{n}{5^n}\right)$$

Using L'Hospital Rule,

$\therefore$
$$L = 1 - \lim_{n\to\infty} \frac{1}{5^n \ln(5)} = 1$$

$\sum b_n = \sum_{n=0}^{\infty} \frac{1}{5^n}$ will converge as it is a geometric series so as $\sum a_n = \sum_{n=0}^{\infty} \frac{3}{5^n - n}$.

EXAMPLE 4.17 Determine the convergence of $\sum \sin^{-1}\left(\frac{1}{n}\right)$.

Ans: Let $\sum b_n = \sum \frac{1}{n}$.

$$L = \lim_{n\to\infty} \frac{a_n}{b_n} = \lim_{n\to\infty} \frac{\sin^{-1}\left(\frac{1}{n}\right)}{1/n}$$

Take $\sin^{-1}\left(\frac{1}{n}\right) = \theta$ we have, $1/n = \sin \theta$,

$$L = \lim_{n\to\infty} \left(\frac{\theta}{\sin \theta}\right) = 1$$

Here, $\sum b_n = \sum \frac{1}{n}$ is divergent series hence $\sum \sin^{-1}\left(\frac{1}{n}\right)$ also diverges.

EXAMPLE 4.18 Check the convergence of $\sum \frac{n^n}{(n+1)^{n+1}}$.

Ans: Let us simplify the given series,

$$a_n = \frac{n^n}{(n+1)^{n+1}} = \frac{n^n}{(n+1)^n (n+1)} = \frac{n^n}{n^n\left(1+\frac{1}{n}\right)^n n\left(1+\frac{1}{n}\right)}$$

$$= \frac{1}{n\left(1+\frac{1}{n}\right)^n\left(1+\frac{1}{n}\right)}$$

From this simplification, if we assume $\sum b_n = \sum \frac{1}{n}$ then we are done.

$$\therefore\ L = \lim_{n\to\infty} \frac{a_n}{b_n} = \lim_{n\to\infty} \frac{1/n\left(1+\frac{1}{n}\right)^n\left(1+\frac{1}{n}\right)}{1/n} = \lim_{n\to\infty} \frac{1}{\left(1+\frac{1}{n}\right)^n\left(1+\frac{1}{n}\right)}$$

$$= \lim_{n\to\infty} \frac{1}{\left(1+\frac{1}{n}\right)^n} = \frac{1}{e}$$

The limit is finite and $\sum b_n = \sum \frac{1}{n}$ is a divergent series hence $\sum \frac{n^n}{(n+1)^{n+1}}$ also diverges.

4.7.4 Ratio Test

Suppose $\sum a_n$ be a series with positive terms, then we define the term $L = \lim_{n\to\infty}\left|\frac{a_{n+1}}{a_n}\right|$. Then there are three possibilities:

1. If $L < 1$ then series is convergent.
2. If $L > 1$ then series is divergent.
3. If $L = 1$ then this test is inconclusive.

EXAMPLE 4.19 Check the convergence of $\sum_{n=1}^{\infty} \frac{(-10)^n}{4^{2n+1}(n+1)}$.

Ans: Using a ratio test, we have to evaluate $L = \lim_{n\to\infty} \left|\frac{a_{n+1}}{a_n}\right|$. For calculating a_{n+1}, we just replace n with $n + 1$.

$$L = \lim_{n\to\infty} \left| \frac{(-10)^{(n+1)}}{4^{2(n+1)+1}((n+1)+1)} \frac{4^{2n+1}(n+1)}{(-10)^n} \right|$$

$$\therefore \qquad L = \lim_{n\to\infty} \left| \frac{(-10)^{(n+1)}}{4^{2n+3}(n+2)} \frac{4^{2n+1}(n+1)}{(-10)^n} \right|$$

$$\therefore \qquad L = \lim_{n\to\infty} \left| \frac{(-10)\,(n+1)}{4^2(n+2)} \right|$$

$$\therefore \qquad L = \frac{10}{16} \lim_{n\to\infty} \left| \frac{(n+1)}{(n+2)} \right|$$

$\therefore\ L = \frac{10}{16} < 1$ hence from the ratio test, we say that the series is convergent.

EXAMPLE 4.20 Determine the convergence of $\frac{3.6.9\ldots 3n.7^n}{4.7.10\ldots(3n+1)(3n+2)}$.

Ans: Finding $L = \lim_{n\to\infty} \left|\frac{a_{n+1}}{a_n}\right|$,

$$a_{n+1} = \frac{3.6.9\ldots 3(n+1)\,.\,7^{n+1}}{4.7.10\ldots(3(n+1)+1)\,(3(n+1)+2)} = \frac{3.6.9\ldots(3n+3)\,.\,7^{n+1}}{4.7.10\ldots(3n+1)\,(3n+4)\,(3n+5)}$$

$$\text{Now, } L = \lim_{n\to\infty} \frac{3.6.9\ldots(3n+3)\,.\,7^{n+1}}{4.7.10\ldots(3n+1)(3n+4)(3n+5)} \times \frac{4.7.10\ldots(3n+1)(3n+2)}{3.6.9\ldots(3n)\,.\,7^n}$$

$$\therefore \qquad L = \lim_{n\to\infty} \frac{(3n+2)(3n+3)7}{(3n+4)(3n+5)} = 7\left(\lim_{n\to\infty} \frac{9n^2(1+2/3n)\,(1+3/3n)}{9n^2(1+4/3n)\,(1+5/3n)} \right) = 7 > 1$$

Hence, the ratio test shows that the given series diverges.

EXAMPLE 4.21 Determine the convergence of $\sum x^n \left(\frac{3n^3 + 7n^2}{5n^9 + 11} \right)$.

Ans: Finding $L = \lim_{n\to\infty} \left|\frac{a_{n+1}}{a_n}\right|$,

$$\therefore \quad L = \lim_{n\to\infty}\left[x^{n+1}\left(\frac{3(n+1)^3+7(n+1)^2}{5(n+1)^9+11}\right)x^n\left(\frac{5n^9+11}{3n^3+7n^2}\right)\right]$$

$$\therefore \quad L = \lim_{n\to\infty}\left[x\left(\frac{3n^3(1+1/n)^3+7n^2(1+1/n)^2}{5n^9(1+1/n)^9+11}\right)\left(\frac{5n^9(1+11/5n^9)}{3n^3(1+7/3n)}\right)\right]$$

$$\therefore \quad L = \lim_{n\to\infty}\left[x\frac{3n^3\left((1+1/n)^3+\frac{7}{3n}(1+1/n)^2\right)}{5n^9\left((1+1/n)^9+\frac{11}{5n^9}\right)}\left(\frac{5n^9(1+11/5n^9)}{3n^3(1+7/3n)}\right)\right]$$

$$\therefore \quad L = \lim_{n\to\infty}\left[x\frac{\left((1+1/n)^3+\frac{7}{3n}(1+1/n)^2\right)}{\left((1+1/n)^9+\frac{11}{5n^9}\right)}\frac{(1+11/5n^9)}{(1+7/3n)}\right]$$

$$= x\frac{((1+0)^3+0(1+0)^2)}{((1+0)^9+0)}\frac{(1+0)}{(1+0)} = x$$

Here, convergence depends on the value of x.

If $x > 1$, the series diverges, converges if $x < 1$ and test fails when $x = 1$.

So, if we simplify the given series,

$$a_n = x^n\frac{3n^3+7n^2}{5n^9+11} = x^n\frac{3n^3(1+7/3n)}{5n^9(1+11/5n^9)} = x^n\frac{3(1+7/3n)}{5n^6(1+11/5n^9)}$$

Then if we assume $b_n = x^n\frac{1}{n^6}$ and calculating $\lim_{n\to\infty}\frac{a_n}{b_n} = \frac{3(1+0)}{5(1+0)} = \frac{3}{5}$.

$\sum b_n = \sum \frac{1}{n^6}$ (p-series test) converges implies the convergence of

$$\sum a_n = \sum x^n\left(\frac{3n^3+7n^2}{5n^9+11}\right)$$

4.7.5 Root Test

Suppose $\sum a_n$ be a series with positive terms, then we define the term $L = \lim_{n\to\infty}\sqrt[n]{|a_n|}$. Then, again this test has three possibilities:

1. If $L < 1$ then series is convergent.
2. If $L > 1$ then series is divergent.
3. If $L = 1$ then this test is inconclusive.

EXAMPLE 4.22 Determine the convergence of $\sum \frac{n^n}{7^{7+3n}}$.

Ans: When the series has power then it is easy to find convergence using root test.

$$\text{Taking } L = \lim_{n\to\infty} \sqrt[n]{|a_n|} = \lim_{n\to\infty} \left(\frac{n^n}{7^{7+3n}}\right)^{1/n}$$

$$\therefore \quad L = \lim_{n\to\infty} \left(\frac{n}{\frac{7}{7^n}+3}\right) = \frac{\infty}{3^2} = \infty > 1 \text{ implies series is divergent.}$$

EXAMPLE 4.23 Determine the convergence of $\sum \sqrt{\frac{n}{n+2}}\, x^n$.

$$\textbf{Ans:} \quad a_n = \sqrt{\frac{n}{n+2}}\, x^n = \left(\frac{n}{n(1+2/n)}\right)^{1/2} x^n = \left(\frac{1}{(1+2/n)}\right)^{1/2} x^n$$

$$\therefore \quad \lim_{n\to\infty} \sqrt[n]{a_n} = \lim_{n\to\infty} \left(\frac{1}{(1+2/n)}\right)^{1/2} x = x$$

From the root test, the given series converges for $|x|<1$ and diverges for $|x|>1$.

When $|x|=1$, $a_n = \sqrt{\frac{n}{n+2}} = \sqrt{\frac{1}{1+2/n}}$

Now, if we apply a comparison test by taking $b_n = \frac{1}{n^0}$.

$$\text{and} \qquad \lim_{n\to\infty} \frac{a_n}{b_n} = \lim_{n\to\infty} \frac{\sqrt{\frac{1}{1+2/n}}}{1/n^0} = 1.$$

As $\sum b_n$ is divergent, given series also diverges for $|x| = 1$.

EXAMPLE 4.24 Check the convergence of $\left[\frac{2^2}{1^2}-\frac{2}{1}\right]^{-1} + \left[\frac{3^3}{2^3}-\frac{3}{2}\right]^{-2} + \left[\frac{4^4}{3^4}-\frac{4}{3}\right]^{-3} + \dots$.

$$\textbf{Ans:} \quad \text{Here, } a_n = \left[\frac{(n+1)^{n+1}}{n^{n+1}} - \frac{n+1}{n}\right]^{-n} = \left(\frac{n+1}{n}\right)^{-n}\left[\left(\frac{(n+1)}{n}\right)^n - 1\right]^{-n}$$

$$\therefore \qquad a_n = \left(1+\frac{1}{n}\right)^{-n}\left[\left(1+\frac{1}{n}\right)^{n}-1\right]^{-n}$$

$$\therefore \qquad a_n^{1/n} = \left(1+\frac{1}{n}\right)^{-1}\left[\left(1+\frac{1}{n}\right)^{n}-1\right]^{-1}$$

$$\therefore \qquad \lim_{n\to\infty} a_n^{1/n} = \lim_{n\to\infty}\left(1+\frac{1}{n}\right)^{-1}\left[\left(1+\frac{1}{n}\right)^{n}-1\right]^{-1} = \left(\frac{1}{1}\right)(e-1)^{-1} = \frac{1}{e-1} < 1$$

Hence, by root test $\sum a_n$ is convergent.

4.8 ABSOLUTELY CONVERGENT AND CONDITIONALLY CONVERGENT SERIES

A series $\sum a_n$ is called **absolutely convergent** if $\sum |a_n|$ is convergent.

If $\sum a_n$ is convergent and $\sum |a_n|$ is divergent we call the series **conditionally convergent.**

EXAMPLE 4.25 Check the convergence of $\sum_{n=1}^{\infty} \frac{(-1)^{n+2}}{n^2}$.

Ans: Let's take $\sum |a_n|$.

$$\sum_{n=1}^{\infty}\left|\frac{(-1)^{n+2}}{n^2}\right| = \sum_{n=1}^{\infty}\frac{1}{n^2}$$

From the p-series test we say that the given series is convergent and hence the series is absolute convergent.

EXAMPLE 4.26 Show that the series $\sum_{n=1}^{\infty}(-1)^n\left(\sqrt{n^2+1}-n\right)$ is conditionally convergent.

Ans: This is an alternating series. Take $|a_n| = \left|(-1)^n\left(\sqrt{n^2+1}-n\right)\right|$

$$\therefore \qquad |a_n| = \left(\sqrt{n^2+1}-n\right)$$

$$\therefore \qquad |a_n| = \frac{\sqrt{n^2+1}-n}{\sqrt{n^2+1}+n}\times\sqrt{n^2+1}+n$$

$$\therefore \qquad |a_n| = \frac{n^2+1-n^2}{\sqrt{n^2+1}+n} = \frac{1}{n\sqrt{1+\frac{1}{n^2}}+1}$$

If we use a comparison test, we can choose $b_n = 1/n$.

Here, $$\lim_{n\to\infty}\frac{|a_n|}{b_n} = \lim_{n\to\infty}\frac{1/\left(n\sqrt{1+\frac{1}{n^2}}+1\right)}{1/n} = \frac{1}{\sqrt{1+\frac{1}{n^2}}+1} = \frac{1}{2}$$

Thus, from comparison test $\sum |a_n|$ is divergent.

$$\lim_{n\to\infty} a_n = \lim_{n\to\infty}\frac{1}{n\sqrt{1+\frac{1}{n^2}}+1} = 0\text{, so } \sum a_n \text{ is convergent.}$$

Therefore, the given series is conditionally convergent.

4.9 POWER SERIES

Till this section, now we have a clear idea about sequence, series, and its convergence. We have also seen some special series and sequences. Now, we will see some applications of the series.

A series of the form $\sum_{n=0}^{\infty} a_n x^n = a_0 + a_1 x + a_2 x^2 + \ldots + a_n x^n + \ldots$ where a_i's are constants called coefficients, is called a power series in x.

A geometric series $\sum_{n=0}^{\infty} ax^n$ is a special power series that consists of the same coefficients of x. If the geometric series $\sum_{n=0}^{\infty} x^n = \frac{1}{1-x}$.

A series of the form $\sum_{n=0}^{\infty} a_n(x-a)^n = a_0 + a_1(x-a) + a_2(x-a)^2 + \ldots + a_n(x-a)^n + \ldots$ where a_i and a are constants, is called a power series in $x - a$. a_i's are called coefficients of the series. The ratio and root tests are two methods to find the radius of convergence. We compute firstly $L = \lim_{n\to\infty}\left|\frac{c^{n+1}}{c^n}\right||x-a|$ or $L = \lim_{n\to\infty}|c_n|^{1/n}|x-a|$. Then the series is convergent for $L < 1$, diverges for $L > 1$ and it will be inclusive for $L = 1$. In this series there exists a number

R such that the power series will converge for $|x-a|<R$, will diverge on $|x-a|>R$ and nothing could be said about on $|x-a|=R$ means series may or may not converge. This number R is called the radius of convergence of the series centered at a. So, the interval of convergence will be, $a - R < x < a + R$.

- The power series may not converge for any value of x, but it always converges for $x = a$. Because,

$$\sum_{n=0}^{\infty} a_n(x-a)^n = \sum_{n=0}^{\infty} a_n(a-a)^n = a_0 0^0 + \sum_{n=1}^{\infty} a_n(0)^n = a_0 .$$

EXAMPLE 4.27 Find the radius of convergence of the series $\sum_{n=0}^{\infty} \frac{(x+2)^n}{3^n}$.

Ans: Here, $\sum_{n=0}^{\infty} \frac{(x+2)^n}{3^n} = \sum_{n=0}^{\infty} \left(\frac{x+2}{3}\right)^n = \frac{1}{1-\frac{x+2}{3}} = \frac{3}{1-x}$

Therefore, $\frac{|x+2|}{3} < 1$

$\therefore \quad |x+2| < 3$

$\therefore \quad -3 < x + 2 < 3$

$\therefore \quad -5 < x < 1$

Thus, the series has the radius of convergence centered at $a = -2$.

EXAMPLE 4.28 Determine the radius of convergence of the series $\sum_{n=0}^{\infty} 4n!(3x+2)^n$

Ans: From the ratio test we have, $L = \lim_{n\to\infty} \left| \frac{4(n+1)!(3x+2)^{n+1}}{4n!(3x+2)^n} \right|$

$\therefore \quad L = \lim_{n\to\infty} \left| \frac{4(n+1)n!(3x+2)^n(3x+2)}{4n!(3x+2)^n} \right|$

$\therefore \quad L = |3x+2| \lim_{n\to\infty} (n+1)$

$\therefore \quad L = \infty > 1$ with $x \neq -(2/3)$

Thus, the series has a radius of convergence centered at $a = -2/3$.

4.10 POWER SERIES AND CALCULUS

In this section, we will see the application of power series. Calculus that deals with derivatives and integration. Now we will see how it applies to power series. Power series is represented as a function and the idea of convergence is important more.

The Geometric series $\sum_{n=0}^{\infty} x^n = \dfrac{1}{1-x}$ provided $|x|<1$. This can be represented as a function $f(x)=\dfrac{1}{1-x}$ provided $|x|<1$. Let's understand with some examples.

- If the power series $f(x)=\sum_{n=0}^{\infty} a_n(x-a)^n$ has a radius of convergence $R > 0$. Then, $f'(x)=\sum_{n=0}^{\infty} na_n(x-a)^{n-1}$ and $\int f(x)\,dx = C + \sum_{n=0}^{\infty} a_n \dfrac{(x-a)^{n+1}}{n+1}$ having a radius of convergence R.

This means that, If power series $f(x)=\sum_{n=0}^{\infty} a_n(x-a)^n$ then,

$$f'(x)=\frac{d}{dx}\sum_{n=0}^{\infty} a_n(x-a)^n = a_1+2a_2(x-a)+3a_3(x-a)^2+\ldots=\sum_{n=1}^{\infty} na_n(x-a)^{n-1}$$

and

$$\int f(x)\,dx=\int \sum_{n=0}^{\infty} a_n(x-a)^n\,dx=\sum_{n=0}^{\infty}\int a_n(x-a)^n\,dx=C+\sum_{n=0}^{\infty} a_n\frac{(x-a)^{n+1}}{n+1}.$$

EXAMPLE 4.29 Determine the power series for $f(x)=\dfrac{2}{1+x^3}$ and then find the interval of convergence of the series.

Ans: Comparing the given function $f(x)=\dfrac{2}{1+x^3}$ with $\sum_{n=0}^{\infty} x^n=\dfrac{1}{1-x}$ we can rewrite the function as, $f(x)=\dfrac{2}{1+x^3}=\dfrac{2}{1-(-x^3)}$ then series will be given by,

$$f(x)=2\sum_{n=0}^{\infty}(-x^3)^n,\ \left|-x^3\right|<1$$

$$\therefore \qquad f(x)=2\sum_{n=0}^{\infty}(-1)^n x^{3n},\ |x|^3<1 \Rightarrow |x|<1$$

So, the interval of convergence is $|x|<1$.

EXAMPLE 4.30 Determine the power series for $f(x)=\dfrac{5x^4}{1+x^3}$ and then find the interval of convergence of the series.

Ans: Comparing the given function $f(x)=\dfrac{5x^4}{1+x^3}$ with $\sum\limits_{n=0}^{\infty} x^n=\dfrac{1}{1-x}$ we can rewrite the function as, $f(x)=\dfrac{5x^4}{1+x^3}=\dfrac{5x^4}{1-(-x^3)}$ then series will be given by,

$$f(x)=5x^4\sum_{n=0}^{\infty}(-x^3)^n,\ \left|-x^3\right|<1,$$

$$\therefore \qquad f(x)=5\sum_{n=0}^{\infty}(-1)^n x^{3n+4},\ |x|^3<1 \Rightarrow |x|<1,$$

So, the interval of convergence is $|x|<1$.

EXAMPLE 4.31 Determine the power series for $f(x)=\dfrac{x^n}{n!}$ and then find the interval of convergence of the series.

Ans: Using the ratio test we have,

$$\left|\frac{u_{n+1}}{u_n}\right|=\left|\frac{x^{n+1}}{(n+1)!}\times\frac{n}{x^n}\right|=\frac{|x|}{n+1}\to 0 \text{ as } n\to\infty \text{ for all value of } x.$$

EXAMPLE 4.32 Find the interval of convergence for the power series $\sum\limits_{n=1}^{\infty}(-1)^{n-1}\dfrac{x^{2n-1}}{2n-1}$.

Ans: Using the ratio test we have,

$$\left|\frac{u_{n+1}}{u_n}\right|=\left|\frac{x^{2(n+1)-1}}{2(n+1)-1}\times\frac{2n-1}{x^{2n-1}}\right|=\left|\frac{x^{2n+1}}{2n+1}\times\frac{2n-1}{x^{2n-1}}\right|=\frac{2n-1}{2n+1}x^2\to x^2 \text{ as } n\to\infty$$

Hence, the series converges if $x^2<1$ and diverges $x^2>1$.

At $x=1$, the given series behaves like an alternating series $1-\dfrac{1}{3}+\dfrac{1}{5}-\dfrac{1}{7}+\ldots$. Hence, it converges.

As at $x=-1$ the series is negative of alternating series $(-1)-\dfrac{(-1)}{3}+\dfrac{(-1)}{5}-\dfrac{(-1)}{7}+\ldots=1-\dfrac{1}{3}+\dfrac{1}{5}-\dfrac{1}{7}+\ldots$, hence it converges.

Thus, the given series is convergent in [–1, 1] and diverges elsewhere.

4.11 TAYLOR AND MACLAURIN SERIES

We know that, how functions are used to represent a series. And Taylor series is a power series which are generated for infinitely many times differentiable functions. Let's take a geometric series having a positive radius of convergence about $x = a$,

$$f(x) = \sum_{n=0}^{\infty} a_n (x-a)^n = a_0 + a_1(x-a) + a_2(x-a)^2 + a_3(x-a)^3 \ldots$$

Taylor Series:

$$f(x) = \sum_{n=0}^{\infty} \frac{f^n(a)}{n!}(x-a)^n = f(a) + f'(a)(x-a) + \frac{f''(a)}{2!}(x-a)^2 + \frac{f'''(a)}{3!}(x-a)^3 + \ldots$$

Maclaurin Series:

$$f(x) = \sum_{n=0}^{\infty} \frac{f^n(0)}{n!}x^n = f(0) + f'(0)x + \frac{f''(0)}{2!}x^2 + \frac{f'''(0)}{3!}x^3 + \ldots$$

EXAMPLE 4.33 Find the Taylor series for $1/x^2$ about $x = -1$.

Ans: By applying the Taylor series formula, we have to compute the successive derivatives.

Let's take, $f(x) = \frac{1}{x^2} \Rightarrow f(-1) = 1$

$$f'(x) = \frac{-2}{x^3} \Rightarrow f'(-1) = \frac{-2}{-1} = 2$$

$$f^{(2)}(x) = \frac{2(3)}{x^4} \Rightarrow f^{(2)}(-1) = \frac{2(3)}{(-1)^4} == 2(3)$$

$$f^{(3)}(x) = -\frac{2(3)(4)}{x^5} \Rightarrow f^{(3)}(-1) = -\frac{2(3)(4)}{(-1)^5} = 2(3)(4)\ldots\ldots\ldots$$

$$f^{(n)}(x) = \frac{(-1)^n(n+1)!}{x^{n+2}} \Rightarrow f^{(n)}(-1) = \frac{(-1)^n(n+1)!}{(-1)^{n+2}} = (n+1)!$$

Hence the Taylor series for the given function is,

$$\frac{1}{x^2} = \sum_{n=0}^{\infty} \frac{f^n(a)}{n!}(x-a)^n = 1 + 2(x+1) + \frac{2(3)}{2!}(x+1)^2 + \frac{2(3)(4)}{3!}(x+1)^3$$

$$+ \ldots + \frac{(n+1)!}{n!}(x+1)^n$$

$$\therefore\ f(x) = \sum_{n=0}^{\infty} (n+1)(x+1)^n$$

EXAMPLE 4.34 Find the Maclaurin series for sin(x).

Ans: Let's take $f(x) = \sin(x)$

$$f(x) = \sin x \Rightarrow f(0) = 0$$

$$\therefore \quad f'(x) = \cos x \Rightarrow f'(0) = 1$$

$$\therefore \quad f''(x) = -\sin x \Rightarrow f''(x) = 0$$

$$\therefore \quad f'''(x) = -\cos x \Rightarrow f'''(x) = -1$$

Thus, the derivatives will be 1, 0, –1, 0, 1, 0, –1, 0, ...

Hence the Maclaurin series,

$$f(x) = \sum_{n=0}^{\infty} \frac{f^n(0)}{n!} x^n = f(0) + f'(0)x + \frac{f''(0)}{2!}x^2 + \frac{f'''(0)}{3!}x^3 + \ldots$$

$$\therefore \sin x = 0 + 1 \cdot x + \frac{0}{2!}x^2 + \frac{(-1)}{3!}x^3 + \frac{0}{4!}x^4 + \frac{1}{5!}x^5 \ldots = x - \frac{x^3}{3!} + \frac{x^5}{5!} - \ldots$$

$$\therefore \sin x = \sum_{n=0}^{\infty} \frac{(-1)^n x^{2n+1}}{(2n+1)!}$$

Note: Similarly, $\cos x = \sum_{n=0}^{\infty} \frac{(-1)^n x^{2n}}{(2n)!}$.

EXAMPLE 4.35 Find the Maclaurin series for x sin (–x).

Ans: As we know the series of sin x, we will use here.

$$\sin x = \sum_{n=0}^{\infty} \frac{(-1)^n x^{2n+1}}{(2n+1)!}$$

$$\text{Hence, } x\sin(-x) = x\sum_{n=0}^{\infty} \frac{(-1)^n(-x)^{2n+1}}{(2n+1)!} = \sum_{n=0}^{\infty} \frac{(-1)^n(-1)^{2n+1}(x)^{2n+1}x}{(2n+1)!}$$

$$= \sum_{n=0}^{\infty} \frac{(-1)^{3n+1}(-x)^{2n+2}}{(2n+1)!}$$

EXAMPLE 4.36 Find the Taylor series for e^x about $x = 2$.

Ans: By applying the Taylor series formula, we have to compute the successive derivatives.

Let's take, $f(x) = e^x$

$f^n(x) = e^x$ implies $f^n(2) = e^2$ for every $n = 0, 1, 2, \ldots..$

Hence the Taylor series for the given function about $x = 2$ is,

$$e^x = \sum_{n=0}^{\infty} \frac{f^n(a)}{n!}(x-a)^n = \sum_{n=0}^{\infty} \frac{e^2}{n!}(x-2)^n$$

EXAMPLE 4.37 Find the Taylor series for $f(x) = \ln(x)$ about $x = 2$.

Ans: By applying the Taylor series formula, we have to compute the successive derivatives.

Let's take, $f(x) = \ln(x)$

$$f(x) = \ln(x) \text{ implies } f(2) = \ln 2$$

$$\therefore \quad f'(x) = \frac{1}{x} \Rightarrow f'(2) = \frac{1}{2}$$

$$\therefore \quad f''(x) = -\frac{1}{x^2} \Rightarrow f''(2) = -\frac{1}{2^2}$$

$$\therefore \quad f'''(x) = \frac{2}{x^3} \Rightarrow f'''(x) = \frac{2}{2^3}$$

$$\therefore \quad f^{iv}(x) = -\frac{2(3)}{x^4} \Rightarrow f^{iv}(x) = -\frac{2(3)}{2^4} \ldots\ldots\ldots$$

$$f^n(x) = \frac{(-1)^{n+1}(n-1)!}{x^n} \Rightarrow f^n(2) = \frac{(-1)^{n+1}(n-1)!}{2^n}$$

Hence, the Taylor series formula can be derived for a given function,

$$\ln(x) = \sum_{n=0}^{\infty} \frac{f^n(a)}{n!}(x-a)^n$$

$$\therefore \quad \ln(x) = f(2) + \sum_{n=1}^{\infty} \frac{f^n(2)}{n!}(x-2)^n = \ln 2 + \sum_{n=1}^{\infty} \frac{(-1)^{n+1}(n-1)!}{n!2^n}(x-2)^n$$

$$\therefore \quad \ln(x) = \ln 2 + \sum_{n=1}^{\infty} \frac{(-1)^{n+1}}{n2^n}(x-2)^n$$

EXAMPLE 4.38 Find the Taylor series for $f(x) = e^x \cos x$.

Ans: Using the Taylor series for $e^x = \sum_{n=0}^{\infty} \frac{x^n}{n!}$ and $\cos x = \sum_{n=0}^{\infty} \frac{(-1)^n x^{2n}}{(2n)!}$

We have, $e^x \cos x = \left(1 + x + \frac{x^2}{2!} + \frac{x^3}{3!} + \frac{x^4}{4!} + \frac{x^5}{5!} + \ldots\right)\left(1 - \frac{x^2}{2!} + \frac{x^4}{4!} - \frac{x^6}{6!} + \ldots\right)$

$$= \left(1+x+\frac{x^2}{2!}+\frac{x^3}{3!}+\frac{x^4}{4!}+\frac{x^5}{5!}+\ldots\right)-\left(\frac{x^2}{2!}+\frac{x^3}{2!}+\frac{x^4}{2!2!}+\frac{x^5}{2!3!}+\ldots\right)$$

$$+\left(\frac{x^4}{4!}+\frac{x^5}{4!}+\frac{x^6}{2!4!}+\ldots\right)+\ldots$$

$$= 1+x-\frac{x^3}{3}-\frac{x^4}{6}-\frac{x^5}{30}+\ldots$$

EXERCISES

Find convergence of sequence:

1. $\left\{\frac{(-1)^{n+1}}{2n-1}\right\}$

2. $\left\{3+(-1)^n\frac{1}{n}\right\}$

3. $\left\{\frac{2}{n^2}\right\}_{n=5}^{\infty}$

4. $\left\{\frac{3-6n^4}{n^4+2n^3}\right\}$

5. $\left\{\frac{(2n+2)!}{(2n-1)!}\right\}$

6. $\left\{\frac{e^{-2n}-2e^{-3n}}{e^{-2n}-e^{-n}}\right\}$

7. $\left\{\left(\frac{3n+1}{3n-1}\right)^n\right\}$

8. $\left\{\frac{(2n)!}{n^n}\right\}$

Find convergence of sequence:

9. $\left(\frac{1}{8}\right)+\left(\frac{1}{8}\right)^2+\left(\frac{1}{8}\right)^3+\left(\frac{1}{8}\right)^4+\ldots$

10. $1+\left(\frac{7}{6}\right)^2+\left(\frac{7}{6}\right)^4+\left(\frac{7}{6}\right)^6+\left(\frac{7}{6}\right)^8+\ldots$

11. $\sum_{n=1}^{\infty}\left(\frac{2}{e}\right)^n(-1)^n$

12. $\sum_{n=1}^{\infty}\frac{2n}{n+4}$

13. $\sum_{n=1}^{\infty}\sin\frac{1}{n}$

14. $\sum_{n=1}^{\infty}\left(\frac{1}{n}-\frac{1}{n+1}\right)$

15. $\sum_{n=0}^{\infty}\left(\frac{5}{2^n}+\frac{1}{3^n}\right)$

16. $\sum\frac{2n!}{(n!)n!}$

17. $\sum_{n=1}^{\infty}\frac{e^n}{n^e}$

18. $\sum\frac{n!\ln n}{(n)(n+2)!}$

19. $\sum \dfrac{\left(n+\sqrt{n}\right)^n}{3^n n^{n+1}}$

20. $\sum_{n=1}^{\infty} \dfrac{\sin n}{n^2}$

21. Find the Maclaurin series for $f(x) = e^x$.

22. Find the Taylor series for $x^3 - 2x + 4$ about $x = 2$.

23. Find the Taylor series for $f(x) = \ln(4+2x)$ about $x = 0$.

24. Find the Maclaurin series for $f(x) = xe^{-x}$.

25. Find the Maclaurin series for $f(x) = e^x \sin x$.

Answers:

1. cgs to 0
2. cgs to 3
3. converges
4. cgs to –6
5. diverges
6. cgs to 0
7. cgs to $e^{2/3}$
8. cgs to $2/e$
9. cgs to $\left(\dfrac{1}{7}\right)$
10. diverges
11. cgs to $\dfrac{e}{e+2}$
12. diverges
13. inconclusive
14. sum = 1, cgs
15. sum = $\dfrac{23}{2}$, cgs
16. diverges
17. diverges
18. converges
19. converges
20. converges
21. $e^x = \sum_{n=0}^{\infty} \dfrac{x^n}{n!}$
22. $8 + (x-2)10 + (x-2)^2.6 + (x-2)^3$
23. $\ln 4 + \dfrac{1}{4}x - \dfrac{1}{6}x^2 + \dfrac{1}{48}x^3 - \ldots$
24. $x - x^2 + \dfrac{x^3}{2!} - \dfrac{x^4}{3!} + \dfrac{x^5}{4!} - \ldots$
25. $x + x^2 + \dfrac{x^3}{3} - \dfrac{x^5}{30} + \ldots$

5 Tangent and Normal

We are now aware of the differentiation of any function. Using the concept of derivative, in this chapter, we define tangent and normal for the curve at a point.

5.1 DEFINITION AND FORMULAE

Tangent to the curve at a given point is the geometry that just touches the curve at that particular point. This geometry can be anything. A circle may be tangent to the curve, and a parabola can be tangent to the curve. But in this chapter, we will consider only tangent lines.

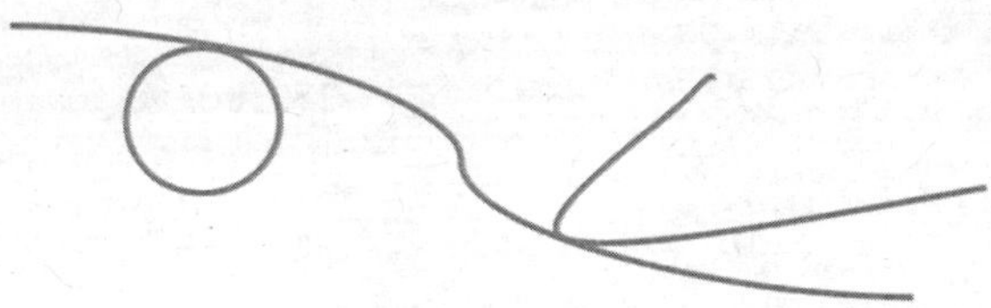

FIGURE 5.1

Any line touching any side of the curve is not necessary to be considered a tangent line. For example, in Figure 5.2(a), line L is not tangent to curve f. In Figure 5.2(b), line L is tangent at point P and intersects at point Q.

Normal to the curve is a line at a point that is perpendicular to the tangent line at that point.

The derivative at a point $P(x_0, y_0)$ defines the slope of the tangent at that point.

i.e., $$\frac{dy}{dx} = f'(x) = \text{slope of tangent} = m(\text{say}) = \tan\,\theta.$$

Then equation of the tangent at a point $P(x_0, y_0)$ is: $y - y_0 = m(x - x_0)$ and the equation of normal at a point $P(x_0, y_0)$ is: $y - y_0 = -1/m(x - x_0)$.

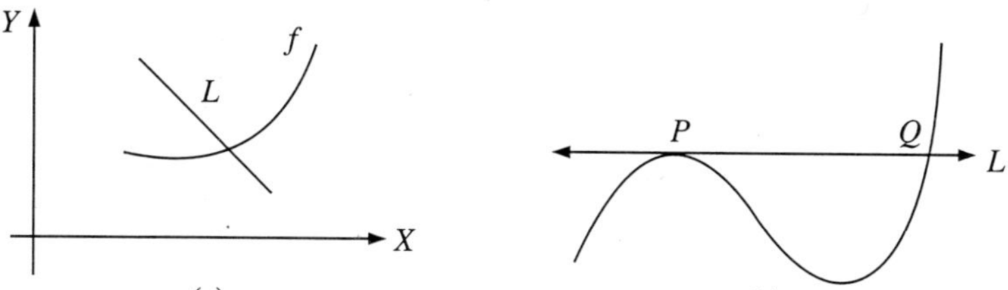

(a) (b)

FIGURE 5.2

It indicates that if we define the slope of tangent (say m_1) and slope of normal (say m_2) at a point then, $m_1 \cdot m_2 = -1$.

Note:

- If the tangent parallel to X–axis at any point, then $\frac{dy}{dx} = m = 0$.
- If tangent parallel to Y–axis at any point, then $\frac{dy}{dx}$ is not defined or $\frac{dx}{dy} = 0$.
- If the tangent on the curve is equally inclined to both axes, $\frac{dy}{dx} = \pm 1$.

Equation of Tangent

1. A curve $y = f(x)$ then the equation of the tangent at any point (x_1, y_1) is, $y - y_1 = m(x - x_1)$, where $m = \frac{dy}{dx} = [f'(x)]_{(x_1, y_1)}$ and equation of normal is, $y - y_1 = -\frac{1}{m}(x - x_1), f'(x)$ exists finitely.
2. If x and y are parametric form that is, $x = f(t)$ and $y = f(t)$. Then the equation of tangent is, $(y - g(t))f'(t) = g'(t)(x - f(t))$, $f'(t) \neq 0$, $f'(t)$ and $g'(t)$ exists finitely.

EXAMPLE 5.1 Find the slope of tangent and normal to the curve. $y = x + \frac{1}{x}$ at $x = 3$. Also, find the equation of tangent.

Ans: Here, $y = x + \frac{1}{x}$. At $x = 3$, $y = 3 + \frac{1}{3} = \frac{10}{3}$.

Hence, the slope of the tangent will be, $\frac{dy}{dx} = m = 1 - \frac{1}{x^2}$ at point $x = 3$ is, $m = 1 - \frac{1}{9} = \frac{8}{9}$.

The slope of normal to the curve is: $\frac{-1}{m} = \frac{-9}{8}$.

Equation of tangent: $y - y_1 = m(x - x_1)$ $y - \frac{10}{3} = \frac{8}{9}(x - 3)$

$$\therefore \quad \frac{8x}{9} - y - \frac{24}{9} + \frac{10}{3} = 0$$

$$\therefore \quad 8x - 9y - 24 + 30 = 0$$

$$\therefore \quad 8x - 9y + 6 = 0$$

EXAMPLE 5.2 Find the slope and equation of tangent to the curve $x = a \cos^3 t$ and $y = a \sin^3 t, t \in \mathbb{R}$.

Ans: Slope of tangent, $\frac{dy}{dx} = \frac{\frac{dy}{dt}}{\frac{dx}{dt}} = \frac{-3a \sin^2 t \cos t}{3a \cos^2 t \sin t} = -\frac{\sin t}{\cos t} = -\tan t$

Hence the equation of the tangent is,

$$y - a \sin^3 t = -\frac{\sin t}{\cos t}(x - a \cos^3 t)$$

$$\Rightarrow \frac{y - a \sin^3 t}{\sin t} = -\frac{(x - a \cos^3 t)}{\cos t}$$

$$\Rightarrow \frac{y}{\sin t} - a \sin^2 t = -\frac{x}{\cos t} + a \cos^2 t$$

$$\Rightarrow \frac{x}{\cos t} + \frac{y}{\sin t} = a \sin^2 t + a \cos^2 t = a$$

$$\Rightarrow x \sec t + y \operatorname{cosec} t = a.$$

EXAMPLE 5.3 Find the points at which tangent to the curve $px^2 + 2kxy + qy^2 = 1$ is perpendicular to the x-axis.

Ans: We have, $px^2 + 2kxy + qy^2 = 1$.

Now, differentiating both sides, $2px + 2kx\frac{dy}{dx} + 2ky + 2qy\frac{dy}{dx} = 0$.

$$\therefore \quad \frac{dy}{dx} = -\frac{px + ky}{kx + qy}$$

The tangent is now perpendicular to the x-axis,

$$\frac{1}{\left(\frac{dy}{dx}\right)} = 0 \Rightarrow kx + qy = 0$$

Hence, the required points are the intersection of the given curve and $kx + qy = 0$.

EXAMPLE 5.4 Show that the points of the curve $y^2 = 4a\left(x + a\sin\frac{x}{a}\right)$ where the tangents are parallel to the x-axis lies on the curve $y^2 = 4ax$.

Ans: Let the curve $y^2 = 4a\left(x + a\sin\frac{x}{a}\right)$ has a tangent at the point (p, q) which satisfies its equation, hence $q^2 = 4a\left(p + a\sin\frac{p}{a}\right)$.

And Tangent $\left(\frac{dy}{dx}\right)_{(p,q)} = 0$ implies $2y\frac{dy}{dx} = 4a\left(1 + \cos\frac{x}{a}\right)$

$$\therefore \quad \left(\frac{dy}{dx}\right)_{(p,q)} = \frac{4a\left(1 + \cos\frac{p}{a}\right)}{2q} = \frac{2a\left(1 + \cos\frac{p}{a}\right)}{q} = 0$$

$$\therefore \quad \cos\frac{p}{a} = -1 \Rightarrow \sin\frac{p}{a} = 0. \text{ This gives } q^2 = 4ap.$$

This proves that the equation of locus for point (p, q) is $y^2 = 4ax$. Hence the proof.

5.2 CARTESIAN SUBTANGENT AND SUBNORMAL

The subtangent is also a line segment associated with the tangent to the curve. Let's consider a curve $y = f(x)$ having tangent line and normal line at point $L(x, y)$ intersecting x-axis at point P and Q respectively. R is the projection of L on x-axis. From Figure 5.3, we have,

- Length of tangent is $LP = |y|\sqrt{1 + \left(\frac{dx}{dy}\right)^2}$
- Length of normal is $LQ = |y|\sqrt{1 + \left(\frac{dy}{dx}\right)^2}$
- Length of subtangent is $PR = |y\cot\theta| = \left|y * \frac{dx}{dy}\right|$
- Length of subnormal is $RQ = |y\cot\theta| = \left|y * \frac{dy}{dx}\right|$.

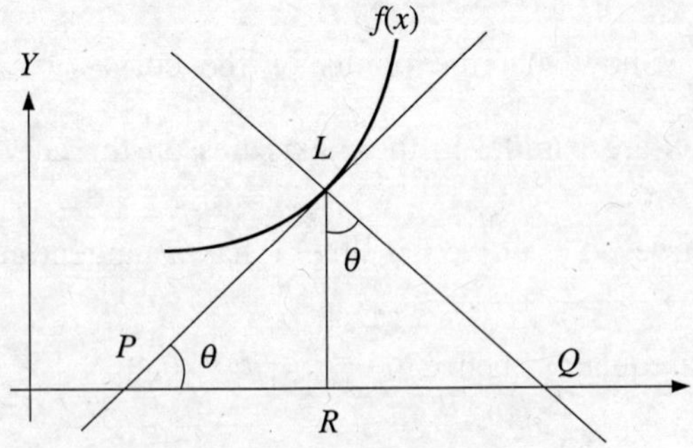

FIGURE 5.3 Cartesian Subtangent and Subnormal.

EXAMPLE 5.5 Determine the lengths of tangent, normal, subtangent, and subnormal to the curve $y^2 = 4ax$ at point $(at^2, 2at)$.

Ans: Here, the curve is $y^2 = 4ax \Rightarrow \dfrac{dy}{dx} = \dfrac{2a}{y}$,

At point, $\left(\dfrac{dy}{dx}\right)_{(at^2,\, 2at)} = \dfrac{2a}{2at} = \dfrac{1}{t}$.

So, the slope of the tangent will be $\left(\dfrac{dy}{dx}\right) = \dfrac{1}{t}$.

$$\text{Length of tangent} = |y|\sqrt{1+\left(\frac{dx}{dy}\right)^2} = 2at\sqrt{1+t^2}$$

$$\text{Length of normal} = |y|\sqrt{1+\left(\frac{dy}{dx}\right)^2} = 2at\sqrt{1+\frac{1}{t^2}} = 2a\sqrt{t^2+1}$$

$$\text{Length of subtangent} = \frac{y}{\left(\frac{dy}{dx}\right)} = \frac{2at}{1/t} = 2at^2$$

$$\text{Length of subnormal} = y\left(\frac{dy}{dx}\right) = 2at \cdot \frac{1}{t} = 2a$$

EXAMPLE 5.6 Determine the lengths of tangent, normal, subtangent, and subnormal to the curve $y = x^2 + x - 1$ at point (1, 1).

Ans: Here, the curve is $y = x^2 + x - 1 \Rightarrow \dfrac{dy}{dx} = 2x + 1$,

$$\Rightarrow \left(\frac{dy}{dx}\right)_{(1,1)} = 3$$

$$\text{Length of tangent} = |y|\sqrt{1+\left(\frac{dx}{dy}\right)^2} = (1)\sqrt{1+\left(\frac{1}{3}\right)^2} = \sqrt{\frac{10}{3}}$$

$$\text{Length of normal} = |y|\sqrt{1+\left(\frac{dy}{dx}\right)^2} = 1\sqrt{1+3^2} = \sqrt{10}$$

$$\text{Length of subtangent} = \frac{y}{\left(\frac{dy}{dx}\right)} = \frac{1}{3}$$

$$\text{Length of subnormal} = y\left(\frac{dy}{dx}\right) = 1\times 3 = 3$$

EXAMPLE 5.7 If the subtangent and subnormal at any point on a curve are equal then what is the length of the tangent?

Ans: Let $M(x_1, y_1)$ be a point on the curve $y = f(x)$. Then the tangent $\frac{dy}{dx} = m$ and the length of the tangent to the curve at this point will be, $|y_1|\sqrt{1+\frac{1}{m^2}}$.

Now, here subtangent and subnormal are equal hence,

$$\frac{y_1}{m} = my_1 \Rightarrow m^2 = 1 \Rightarrow m = \pm 1$$

And the length of the tangent will be, $|y_1|\sqrt{1+\frac{1}{1}} = |y_1|\sqrt{2}$.

5.3 POLAR CO-ORDINATES

Take a fixed origin O and two perpendicular lines passing through the origin called axes. For Cartesian co-ordinates of point $P(x, y)$, displacement in the x-direction say x, and displacement in the y-direction says y. In a polar coordinate system, a distance from O to P in radius direction say r and an angle made with polar line (initial line) OL is θ. In an anticlockwise direction. Now, the point has polar coordinates $P(r, \theta)$ and one can say $r = f(\theta)$.

Here, always $r > 0$ and $-\pi < \theta < \pi$ is measured in radians.

Coordinate Conversion

From Figure 5.4, we can derive the formula for converting Cartesian to polar coordinates.

$$\cos\theta = \frac{x}{r} \Rightarrow x = r\cos\theta \text{ and } \sin\theta = \frac{y}{r} \Rightarrow y = r\sin\theta$$

Also, $$x^2 + y^2 = r^2 \text{ and } \tan\theta = \frac{y}{x} \Rightarrow \theta = \tan^{-1}\frac{y}{x}.$$

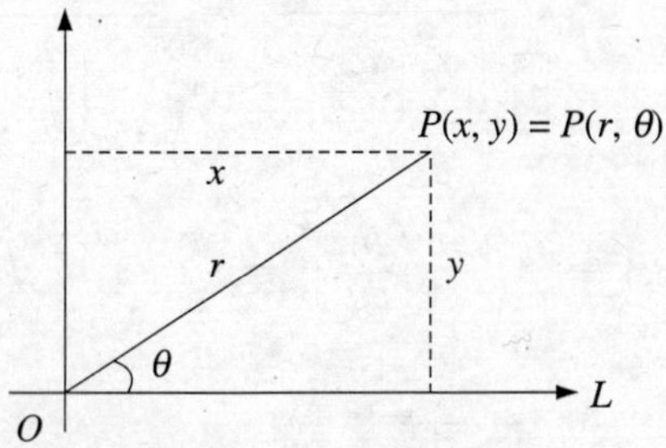

FIGURE 5.4 Polar Representation.

5.4 ANGLE BETWEEN RADIUS AND TANGENT

The angle between the radius and tangent of a circle is a fundamental concept in geometry, particularly in understanding the relationships between points on a circle and its tangent line. Let's draw a circle having center *O*.

Figure 5.5 helps to find the angle between radius and tangent. Here, *OS* is the radius and *OQ* is the tangent for the circle at point *S*. At the point where the tangent touches the circle (point of tangency), the tangent is perpendicular to the radius. In geometric terms, this angle is always 90 degrees (or $\pi/2$ radians). This means that the radius and tangent form a right angle where they meet on the circumference of the circle.

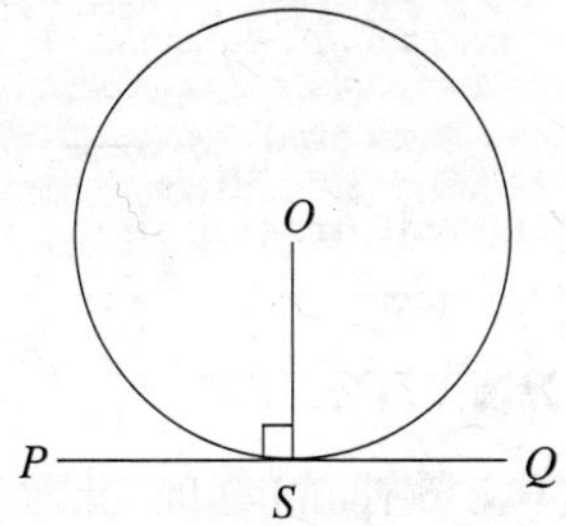

FIGURE 5.5 Angle between Radius and Tangent.

5.5 PERPENDICULAR LENGTH FROM POLE ON TANGENT

Consider a pole *O* having line *OL* and curve $r = f(\theta)$. Draw a tangent at the point *P*.

Let the angle between the radius vector and the line is θ and the angle between the radius vector and the tangent line is ϕ.

Now, consider the perpendicular line *OR* from pole *O* to tangent forming a right-angle triangle ΔONP,

$$\sin\phi = \frac{opp}{hyp} = \frac{ON}{OP} = \frac{p}{r}$$

$$\boxed{r\sin\phi = p} \tag{5.1}$$

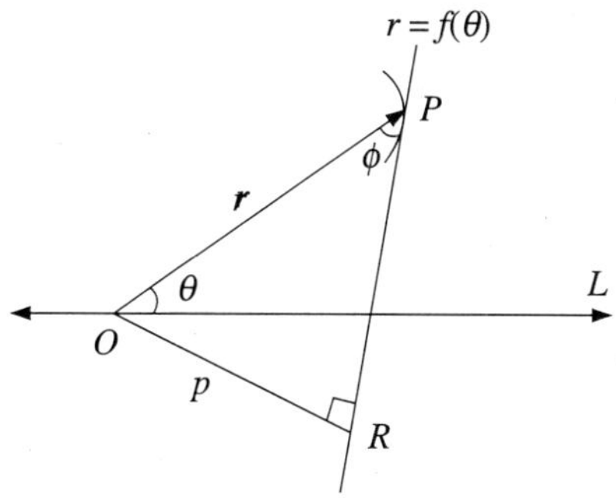

FIGURE 5.6 Perpendicular Length.

Now, taking reciprocal $\dfrac{1}{r\sin\phi} = \dfrac{1}{p} \Rightarrow \dfrac{1}{p} = \dfrac{1}{r}\operatorname{cosec}\phi$. Squaring we have, $\dfrac{1}{p^2} = \dfrac{1}{r^2}\operatorname{cosec}^2\phi = \dfrac{1}{r^2}(1+\cot^2\phi)$.

But, $\boxed{\tan\phi = r\cdot\dfrac{d\theta}{dr}} \Rightarrow \cot\phi = \dfrac{1}{r}\cdot\dfrac{dr}{d\theta}$.

So,
$$\frac{1}{p^2} = \frac{1}{r^2}\left(1+\left(\frac{1}{r}\frac{dr}{d\theta}\right)^2\right) = \frac{1}{r^2}\left(1+\frac{1}{r^2}\left(\frac{dr}{d\theta}\right)^2\right)$$

∴
$$\boxed{\frac{1}{p^2} = \frac{1}{r^2} + \frac{1}{r^4}\left(\frac{dr}{d\theta}\right)^2} \tag{5.2}$$

Let $\dfrac{1}{r} = u$.

Differentiating we have, $-\dfrac{1}{r^2}\left(\dfrac{dr}{d\theta}\right) = \dfrac{du}{d\theta} \Rightarrow -\dfrac{1}{r^4}\left(\dfrac{dr}{d\theta}\right)^2 = \left(\dfrac{du}{d\theta}\right)^2$

Substituting in the above Equation (5.2),

∴
$$\boxed{\frac{1}{p^2} = u^2 + \left(\frac{du}{d\theta}\right)^2} \tag{5.3}$$

5.6 PEDAL EQUATION

The pedal equation of curve defined the relation between perpendicular distance from pole on tangent i.e., p and radius vector from the pole i.e., r.

Let's consider the Cartesian equation of the curve defined in Figure 5.6, given by $y = f(x)$.

Equation of tangent at any point (x, y) is,

$$Y - y = \frac{dy}{dx}(X - x) = X\frac{dy}{dx} - x\frac{dy}{dx}$$

$$\therefore \qquad p = \frac{y - x\dfrac{dy}{dx}}{\sqrt{1 + \left(\dfrac{dy}{dx}\right)^2}} \tag{5.4}$$

Pedal coordinates are, $r = \sqrt{x^2 + y^2}$. (5.5)

The equation of the curve will be, $f(x, y) = 0$. The pedal equation can be obtained by eliminating (x, y) from equation of curve, equation (5.4) and (5.5).

The equation of the tangent at any point (x, y) on the curve is, $xf_x + yf_y - (xf_x + yf_y) = 0$.

$$p = \frac{x\dfrac{\partial f}{\partial x} + y\dfrac{\partial f}{\partial y}}{\sqrt{\left(\dfrac{\partial f}{\partial x}\right)^2 + \left(\dfrac{\partial f}{\partial y}\right)^2}}$$

EXAMPLE 5.8 Find the pedal equation for the polar curve $\dfrac{2a}{r} = 1 - \cos\theta$.

Ans: Here, $r = \dfrac{2a}{1 - \cos\theta} = \dfrac{2a}{2\sin^2\dfrac{\theta}{2}} = a\operatorname{cosec}^2\dfrac{\theta}{2}$

$$\frac{dr}{d\theta} = 2a\operatorname{cosec}\frac{\theta}{2}\cdot\frac{1}{2}\left(-\operatorname{cosec}\frac{\theta}{2}\cot\frac{\theta}{2}\right) = -a\operatorname{cosec}^2\frac{\theta}{2}\cot\frac{\theta}{2}$$

Now, we know that

$$\tan\phi = r\cdot\frac{d\theta}{dr} = \frac{r}{\left(\dfrac{dr}{d\theta}\right)} = \frac{a\operatorname{cosec}^2\dfrac{\theta}{2}}{-a\operatorname{cosec}\dfrac{\theta}{2}\cot\dfrac{\theta}{2}} = -\tan\frac{\theta}{2} = \tan\left(\pi - \frac{\theta}{2}\right)$$

Hence, $\phi = \left(\pi - \dfrac{\theta}{2}\right)$

Pedal equation, $p = r\sin\phi = r\sin\left(\pi - \dfrac{\theta}{2}\right) = r\sin\dfrac{\theta}{2}$

Squaring we get, $p^2 = r^2\sin^2\dfrac{\theta}{2}$.

$$\frac{p^2}{r^2} = \sin^2\frac{\theta}{2} \Rightarrow \operatorname{cosec}^2\frac{\theta}{2} = \frac{r^2}{p^2}$$

Initially, we have, $r = a\operatorname{cosec}^2\frac{\theta}{2} \Rightarrow \frac{r}{a} = \operatorname{cosec}^2\frac{\theta}{2}$

So, comparing the above two expressions we have, $\frac{r}{a} = \frac{r^2}{p^2}$.

$p^2 = ar$ is a required equation.

EXAMPLE 5.9 Find the pedal equation for the polar curve $r^m = a^m \sin m\theta + b^m \cos m\theta$.

Ans: Consider, the given equation $r^m = a^m \sin m\theta + b^m \cos m\theta$.

Taking log on both sides we have, $\log r^m = \log(a^m \sin m\theta + b^m \cos m\theta)$

$$m\frac{1}{r}\frac{dr}{d\theta} = \frac{m(a^m \cos m\theta - b^m \sin m\theta)}{a^m \sin m\theta + b^m \cos m\theta} \Rightarrow \frac{1}{r}\frac{dr}{d\theta} = \frac{(a^m \cos m\theta - b^m \sin m\theta)}{a^m \sin m\theta + b^m \cos m\theta}$$

As we have, $\tan\phi = r\cdot\frac{d\theta}{dr}$

$$\cot\phi = \frac{(a^m \cos m\theta - b^m \sin m\theta)}{a^m \sin m\theta + b^m \cos m\theta}$$

Also, $p = r\sin\phi \Rightarrow \frac{1}{p} = \frac{1}{r}\operatorname{cosec}\phi$

$$\Rightarrow \frac{1}{p^2} = \frac{1}{r^2}\operatorname{cosec}^2\phi = \frac{1}{r^2}(1+\cot^2\phi)$$

$$\Rightarrow \frac{1}{p^2} = \frac{1}{r^2}\left(1+\left(\frac{(a^m \cos m\theta - b^m \sin m\theta)}{a^m \sin m\theta + b^m \cos m\theta}\right)^2\right)$$

$$\Rightarrow \frac{1}{p^2} = \frac{1}{r^2}\left(\frac{(a^m \sin m\theta + b^m \cos m\theta)^2 + (a^m \cos m\theta - b^m \sin m\theta)^2}{(a^m \sin m\theta + b^m \cos m\theta)^2}\right)$$

After simplifying,

$$\Rightarrow \frac{1}{p^2} = \frac{1}{r^2}\left(\frac{a^{2m}+b^{2m}}{r^{2m}}\right)$$

Hence, the required pedal equation is, $\Rightarrow p^2 = \frac{r^{2m+2}}{a^{2m}+b^{2m}}$.

5.7 ARC IN CARTESIAN FORM

From Figure 5.7(a), let's take a curve $y = f(x)$ having points P and Q of distance $r = q - p$. By using the theory of closer and closer approximations, we approximate the length of the arc by adding line segments between points P and Q. If we increase the number of line segments, it gives a better approximation.

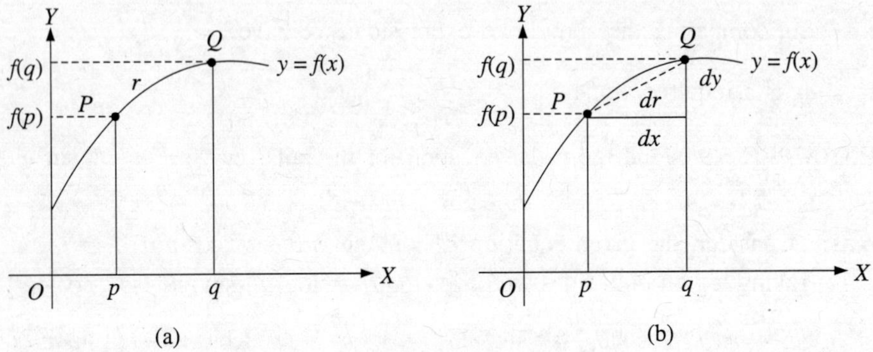

FIGURE 5.7

Choose one of the segments denoted by dr and the corresponding change with respect to x and y direction is denoted by dx and dy.

Now, from Figure 5.7(b) we have,

$$dr = \sqrt{(dx)^2 + (dy)^2}$$

Now, as we mentioned above the length of the curve is the sum of line segments

$$r = \lim_{dx \to 0} \sum dr$$

Hence, $r = \lim\limits_{dx \to 0} \sum\limits_{p}^{q} \sqrt{(dx)^2 + (dy)^2} = \int\limits_{p}^{q} \sqrt{(dx)^2 + (dy)^2}$

Trying to convert in a simplified form that is,

$$r = \int_{p}^{q} \sqrt{(dx)^2 \left(1 + \left(\frac{dy}{dx}\right)^2\right)} = \int_{p}^{q} \sqrt{\left(1 + \left(\frac{dy}{dx}\right)^2\right)}\, dx$$

A similar formula with respect to y we can have, $r = \int\limits_{p}^{q} \sqrt{\left(1 + \left(\frac{dx}{dy}\right)^2\right)}\, dy$.

EXAMPLE 5.10 Determine the length of the curve $x^2 = 4y$ from $x = 0$ to $x = 2$.

Ans: Here, function is $x^2 = 4y \Rightarrow y = \dfrac{x^2}{4}$.

Hence, $\dfrac{dy}{dx} = \dfrac{x}{2}$

$$r = \int_0^2 \sqrt{\left(1 + \left(\frac{x}{2}\right)^2\right)}\, dx = \int_0^2 \frac{1}{2}\sqrt{(4 + x^2)}$$

$$\therefore \qquad r = \frac{1}{2}\left[x\sqrt{(4+x^2)} + 2\ln\left(\frac{x + \sqrt{(4+x^2)}}{2}\right)\right]_0^2$$

$$\therefore \qquad r = \frac{1}{2}\left[(2\sqrt{8} - 0) + 2\left(\ln\left(\frac{2+\sqrt{8}}{2}\right) - \ln 1\right)\right]$$

$$r = 229.$$

5.8 ARC IN POLAR FORM

The parametric representation helps us to find the arc length. The arc length of the parameterized curve is given by, $L = \int_p^q \sqrt{\left(\frac{dx}{d\theta}\right)^2 + \left(\frac{dy}{d\theta}\right)^2}\, d\theta$.

As discussed in section 5.3, $r = f(\theta)$ having co-ordinates,

$$x = f(\theta) \cos\theta = r\cos\theta$$

$$y = f(\theta) \sin\theta = r\sin\theta$$

Now by evaluating derivatives,

$$\frac{dx}{d\theta} = \frac{dr}{d\theta}\cos\theta - r\sin\theta \quad \text{and} \quad \frac{dy}{d\theta} = \frac{dr}{d\theta}\sin\theta + r\cos\theta$$

Now, by squaring the above expression we have

$$\left(\frac{dx}{d\theta}\right)^2 = \left(\frac{dr}{d\theta}\right)^2 \cos^2\theta + r^2\sin^2\theta - 2r\frac{dr}{d\theta}\sin\theta\cos\theta$$

$$\left(\frac{dy}{d\theta}\right)^2 = \left(\frac{dr}{d\theta}\right)^2 \sin^2\theta + r^2\cos^2\theta + 2r\frac{dr}{d\theta}\sin\theta\cos\theta$$

Adding these two gives,

$$\left(\frac{dx}{d\theta}\right)^2 + \left(\frac{dy}{d\theta}\right)^2 = \left(\frac{dr}{d\theta}\right)^2 (\sin^2\theta + \cos^2\theta) + r^2(\sin^2\theta + \cos^2\theta) = r^2 + \left(\frac{dr}{d\theta}\right)^2$$

So, $L = \int_p^q \sqrt{\left(\frac{dx}{d\theta}\right)^2 + \left(\frac{dy}{d\theta}\right)^2}\, d\theta = \int_p^q \sqrt{r^2 + \left(\frac{dr}{d\theta}\right)^2}\, d\theta = \int_p^q \sqrt{(f(\theta))^2 + (f'(\theta))^2}\, d\theta$

EXAMPLE 5.11 Determine the length of the curve $r = 2 + 2\cos\theta$.

Ans: As the given curve is cardioid, so when θ goes from 0 to 2π, it traced out exactly once.

Therefore,

$$L = \int_p^q \sqrt{(f(\theta))^2 + (f'(\theta))^2}\, d\theta$$

$$= \int_0^{2\pi} \sqrt{(2 + 2\cos\theta)^2 + (-2\sin\theta)^2}\, d\theta$$

$$= \int_0^{2\pi} \sqrt{4 + 8\cos\theta + 4\cos^2\theta + 4\sin^2\theta}\, d\theta$$

$$= \int_0^{2\pi} \sqrt{8 + 8\cos\theta}\, d\theta$$

$$= 2\int_0^{2\pi} \sqrt{2 + 2\cos\theta}\, d\theta$$

$$= 2\int_0^{2\pi} \sqrt{4\cos^2\left(\frac{\theta}{2}\right)}\, d\theta = 4\int_0^{2\pi} \left|\cos\left(\frac{\theta}{2}\right)\right| d\theta$$

This can be written as, $L = 8\int_0^{\pi} \cos\left(\frac{\theta}{2}\right) d\theta = 16$.

EXERCISES

Evaluate given problems:

1. Find the point at which the curve has having slope of tangent 3/8.
2. Find the equation of tangent to the curve $y = (x^3 - 1)(x - 2)$ at the points where the curve cuts the x-axis.
3. Find the normal to the curve $x + y = x^y$ which cuts the x-axis.
4. The curve $x = a(\theta + \sin\theta)$ and $y = a(1 - \cos\theta)$, $a \neq 1$ I defined. If PQ and PR are the lengths of subtangent and subnormal on the given curve at $\theta = \pi/2$, what is the relation between PQ and PR.
5. Prove that the length of the subtangent and subnormal of the curve $y = be^{x/a}$ is constant.
6. Find the length of the subnormal of the curve $y = \sin x$ at point $x = \pi/4$.
7. What is the pedal equation for $r = e^{\theta \cot \alpha}$?
8. Find the pedal equation for $r^2 = a^2 \sec 2\theta$.
9. Find a pedal equation for $r = a(1 + \cos\theta)$
10. Find a pedal equation for $x = ae^{\theta}(\sin\theta - \cos\theta)$ and $y = ae^{\theta}(\sin\theta + \cos\theta)$.
11. Find the arc length between 0 to 1.
12. Find the arc length over the interval $[0, \pi]$.
13. Find the arc length of $x = 3 \sin 3t$ and $y = 3 \cos 3t$, $0 \leq t \leq 2\pi$.
14. Find the length of the curve $r = \theta$, $0 \leq \theta \leq 1$.
15. Find the length of the curve $r = 3 \sin\theta$.

Answers:

1. (2, –2)	2. $7x - y - 14 = 0$	3. $y = x - 1$
4. PQ = PR	5. constant $= a$	6. 0.5
7. $p = r \sin\alpha$	8. $p = a^2/r$	9. $2ap^2 = r^3$
10. $2p^2 = r^2$	11. 2.268 units	12. 3.82 units
13. 18π	14. $\frac{1}{2}(\sqrt{2} + \ln(1 + \sqrt{2}))$	15. 3π

6 Partial Differentiation

In previous classes, we have learnt about a function having one independent variable (say) x and one dependent variable (say) y. It is denoted by $y = f(x)$. For example, volume of the sphere depends only on radius and we write $V = \frac{4}{3}\pi r^3$.

6.1 FUNCTION OF SEVERAL VARIABLES

In this section, we will discuss function of n-independent variables and one dependent variable. Let $X = \{x_1, x_2, ..., x_n\}$ be a set of n-tuples of real numbers. A real value of function f on x assigns a unique real number $u = f\{x_1, x_2, ..., x_n\}$ to each element of X. The set X is called domain of the function. The set of values u taken by f is called the range of the function. u is dependent variable and $x_1, x_2, ..., x_n$ are n-independent variables of the functions. In electrical engineering, n-independent variables are known as inputs and a dependent variable is output of an experiment. In chemical engineering, n-independent variables are n-chemical compounds and a dependent variable is final chemical obtained by chemical reactions. In pollution science, air, water and plant pollutions are independent variables to create environment pollution which is dependent variable. The volume of the cylinder is $\pi r^2 h$ which has two independent variables, viz radius, r of the cylinder and height h of the cylinder. In general, if f is a function of two independent variables (say) x and (say) y and dependent variable u, we write $u = f(x, y)$. Similarly, $u = f(x, y, z)$ denotes function of three variables. ***The domain of a function is the set for which the defined rule generates real numbers. The range is the set of output values for the dependent variable.***

Let us see the domain and range of function of two variables.

	Function	*Domain*	*Range*
1.	$u = \sqrt{x^2 - y}$	$x^2 \geq y$	[0, 2)

2. $u = \dfrac{4+x}{xy}$ $\quad xy \neq 0 \quad (-\alpha, 0) \cup (0, 2)$

3. $u = \cos xy$ $\quad \mathbb{R}^2$ Entire $\quad [-1, 1]$

Similarly, one can find domain and range for a function of three variables.

A point (x_0, y_0) in xy–plane is said to be interior point of S if there is disk centered around (x_0, y_0) having a positive radius that lies entirely in S as shown in Figure 6.1.

A point (x_0, y_0) is said to be boundary Point of S if every disk centered at (x_0, y_0) contain points lying outside S as well as lying inside S. It is shown in Figure 6.2.

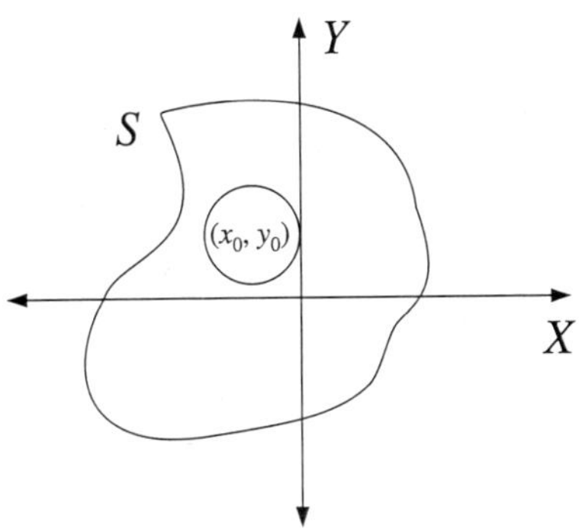

FIGURE 6.1

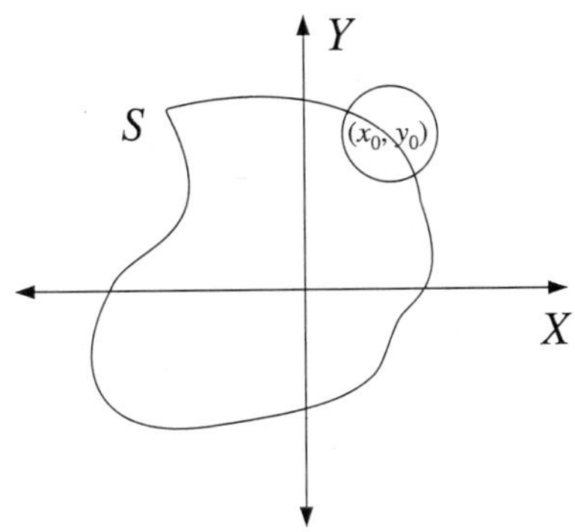

FIGURE 6.2

A region in the plane is bounded if it lies inside a disk of fixed radius. A region which is not bounded is said to be unbounded. Line segment, triangle, circle, rectangle and disk are examples of bounded region whereas graph of function defined on infinite interval, quadrant, half-plane and plane are examples of an unbounded region.

The set of points in the plane for which $f(x, y) = c$ takes constant value is called a level curve of f. The set of all points $(x, y, f(x, y))$ in space is called the graph of f or surface $u = f(x, y)$.

6.2 LIMIT AND CONTINUITY

A function $f(x, y)$ tends to limit L as $(x, y) \to (x_0, y_0)$ and we write

$$\lim_{(x,y)\to\infty(x_0,y_0)} f(x, y) = L$$

if for given $\varepsilon > 0$, there exist $\delta > 0$ such that for all (x, y) in the domain of f,

$$|f(x, y) - L| < \varepsilon, \text{ whenever } 0 < \sqrt{(x-x_0)^2 + (y-y_0)^2} < \delta.$$

i.e., the distance between $f(x, y)$ and L becomes arbitrary small whenever distance between (x, y) and (x_0, y_0) reduces to very small quantity. This concept is true for interior points as well as boundary points.

We have seen that for a function of a single variable,

$$\lim_{(x,y)\to(x_0,y_0)} x = x_0$$

$$\lim_{(x,y)\to(x_0,y_0)} y = y_0$$

$$\lim_{(x,y)\to(x_0,y_0)} c = c, \text{ } c \text{ is scalar.}$$

In the first limit statement, $f(x, y) = x$ and $L = x_0$. Let $\varepsilon > 0$ be given. Set $\delta = \in$. Then

$$0 < \sqrt{(x-x_0)^2 + (y-y_0)^2} < \delta = \varepsilon$$

implies $$\sqrt{(x-x_0)^2} < \in$$

$$\Rightarrow |x - x_0| < \varepsilon$$

$$\Rightarrow |f(x,y) - x_0| < \varepsilon$$

i.e., $|f(x,y) - x_0| < \varepsilon < \in$, whenever $0 < \sqrt{(x-x_0)^2 + (y-y_0)^2} < \delta$.

Properties of Limits of Function of Two Variables

Let m, n and c be real numbers.

Given $\lim_{(x,y)\to(x_0,y_0)} f(x,y) = m$ and $\lim_{(x,y)\to(x_0,y_0)} g(x,y) = n$.

1. Sum Rule: $\lim_{(x,y)\to(x_0,y_0)} (f(x,y) + g(x,y)) = m + n$.
2. Difference Rule: $\lim_{(x,y)\to(x_0,y_0)} (f(x,y) - g(x,y)) = m - n$.
3. Scalar Multiplication Rule: $\lim_{(x,y)\to(x_0,y_0)} cf(x,y) = cm$.
4. Product Rule: $\lim_{(x,y)\to(x_0,y_0)} (f(x,y)\, g(x,y)) = mn$.
5. Quotiet Rule: $\lim_{(x,y)\to(x_0,y_0)} \dfrac{f(x,y)}{g(x-y)} = \dfrac{m}{n}, n \neq 0$.
6. Power Rule: $\lim_{(x,y)\to(x_0,y_0)} [f(x,y)]^k = m^k$, k is a positive integer.

EXAMPLE 6.1 Find $\lim_{(x,y)\to(0,\,0)} \dfrac{x + xy - 3}{x^2 + xy + 6}$.

Ans: $$\lim_{(0,\,0)\to(0,\,0)} \frac{x + xy - 3}{x^2 + xy + 6} = \frac{-3}{6} = \frac{-1}{2}.$$

EXAMPLE 6.2 Find $\lim\limits_{(x,y)\to(-3,\,4)} \dfrac{\sqrt{x^2+y^2}}{x}$.

Ans: $\lim\limits_{(x,y)\to(-3,\,4)} \dfrac{\sqrt{x^2+y^2}}{x} = \dfrac{\sqrt{(-3)^2+4^2}}{-3} = -\dfrac{5}{3}$.

In these two examples, we have simply substituted values of x and y.

EXAMPLE 6.3 Find $\lim\limits_{(x,y)\to(1,1)} \dfrac{x^3-x^2y}{\sqrt{x}-\sqrt{y}}$.

Ans: $\lim\limits_{(x,y)\to(1,1)} \dfrac{x^3-x^2y}{\sqrt{x}-\sqrt{y}} = \lim\limits_{(x,y)\to(1,1)} \dfrac{x^2(\sqrt{x}+\sqrt{y})(\sqrt{x}-\sqrt{y})}{\sqrt{x}-\sqrt{y}} = \lim\limits_{(x,y)\to(1,1)} x^2(\sqrt{x}+\sqrt{y}) = 2.$

EXAMPLE 6.4 Find $\lim\limits_{(x,y)\to(0,0)} \dfrac{xy^2}{x^2+y^2}$ (if it exists).

Ans: We can see that along the line $x = 0$, value of the function is zero when $y \neq 0$. Similar is the behavior along the line $y = 0$ when $x \neq 0$. This suggests that if the limit exists, then the value of the limit must be zero.

Given $\varepsilon > 0$. We find $\delta > 0$ such that $\left|\dfrac{xy^2}{x^2+y^2} - 0\right| < \varepsilon$ whenever $0 < \sqrt{x^2+y^2} < \delta$.

i.e. $\dfrac{|x|y^2}{x^2+y^2} < \varepsilon$ whenever $0 < \sqrt{x^2+y^2} < \delta$ since $y^2 < x^2 + y^2$. We have $\dfrac{|x|y^2}{x^2+y^2} \le |x| = \sqrt{x^2} \le \sqrt{x^2+y^2}$.

This suggests that choose $\delta = \varepsilon$ and let $0 < \sqrt{x^2+y^2} < \delta$, we get

$$\left|\frac{xy^2}{x^2+y^2} - 0\right| \le \sqrt{x^2+y^2} < \delta = \varepsilon$$

Hence, it follows from the definition of the limit that $\lim\limits_{(x,y)\to(x_0,y_0)} \dfrac{xy^2}{x^2+y^2} = 0$.

Similar to function of a single variable continuity in terms of limits is defined as follows:

A function $f(x, y)$ is continuous at a point (x_0, y_0) if

1. f is defined at (x_0, y_0),
2. $\lim\limits_{(x,y)\to(x_0,y_0)} f(x,y)$ exists, and
3. $\lim\limits_{(x,y)\to(x_0,y_0)} f(x,y) = f(x_0,y_0)$.

A function is continuous if it is continuous at every point of its domain.

Using properties of limit of functions, the algebraic combinations of continuous functions are continuous at every point at which all the functions are well defined.

EXAMPLE 6.5 Show that $f(x, y) = \begin{cases} \dfrac{xy}{x^2 + y^2}, & (x, y) \neq (0, 0) \\ 0 & , (x, y) = (0, 0) \end{cases}$ is continuous at every point except at (0, 0).

Ans: For $(x, y) \neq (0, 0)$, the function is continuous because its values are given by a rational function of x and y and limit value can be obtained by direct substitution of values of x and y.

At (0, 0), $f(0, 0)$ is defined and its value is zero. Let us assume that f has no limit when $(x, y) \to (0, 0)$. We will try to prove this by choosing different paths leading to different limits at (0, 0).

Consider, the path $y = mx$, $x \neq 0$. Then $\lim\limits_{(x,y)\to(0,0)} \dfrac{xy}{x^2 + y^2} = \dfrac{x(mx)}{x^2 + m^2x^2}$

$$= \frac{m}{1 + m^2}.$$

This suggests that the limit of the function is dependent on m. So, function is not continuous.

EXAMPLE 6.6 Show that $\lim\limits_{(x,y)\to(0,0)} \dfrac{y^2 - x^2}{y^2 + x^2}$ does not exist.

Ans: Check that if $y = 0$ and $x \neq 0$ then $f(x, y) = -1$, whereas $x = 0$ and $y \neq 0$ gives $f(x, y) = 1$, Therefore, limit does not exist at (0, 0).

EXAMPLE 6.7 Evaluate $\lim\limits_{(x,y)\to(0,0)} \dfrac{x^2y}{x^4 + y^2}$.

Ans: $\lim\limits_{\substack{x\to 0 \\ y\to 0}} \dfrac{x^2y}{x^4 + y^2} = \lim\limits_{y\to\infty} 0 = 0$ and $\lim\limits_{\substack{y\to 0 \\ x\to 0}} \dfrac{x^2y}{x^4 + y^2} = \lim\limits_{x\to\infty} 0 = 0.$

$$\lim_{\substack{y\to mx \\ x\to 0}} \frac{x^2y}{x^4 + y^2} = \lim_{x\to 0} \frac{mx^3}{x^4 + m^2x^2} = \lim_{x\to 0} \frac{mx}{x^2 + m^2} = 0.$$

$$\lim_{\substack{y\to mx^2 \\ x\to 0}} \frac{x^2y}{x^4 + y^2} = \lim_{x\to 0} \frac{mx^4}{x^4 + m^2x^4} = \frac{m}{1 + m^2}.$$

Here, the limit along the path $y = mx^2$ depends on m, therefore limit does not exist.

EXAMPLE 6.8 $\lim\limits_{(x,y)\to(e,1)} \ln\left(\frac{x}{y}\right)$.

Ans: Let $f(x,y) = \frac{x}{y}$ and $g(u) = \ln u$.

Using quotient rule for limits, $\lim\limits_{(x,y)\to(e,1)} f(x,y) = \frac{\lim\limits_{(x,y)\to(e,1)} x}{\lim\limits_{(x,y)\to(e,1)} y} = \frac{e}{1} = e$

$g(u) = \ln u$ is continuous at e and $g(e) = \ln e = 1$. So, $\lim\limits_{(x,y)\to(e,1)} \ln\left(\frac{x}{y}\right) = \ln\left(\frac{e}{1}\right)$

$= \ln e = 1$.

Two-Path Test Form nonexistence of a Limit

If a function $f(x, y)$ has different limits along different paths in the domain of a function f as $(x, y) \to (x_0, y_0)$, then $\lim\limits_{(x,y)\to(x_0,y_0)} f(x,y)$ does not exist.

6.3 PARTIAL DERIVATIVE OF A FUNCTION

Let $u = f(x, y)$. The derivative of u with respect to x (if it exists) keeping y constant is called the partial derivative of u with respect to x and is denoted by $\partial u/\partial x$ or u_x and defined as:

$$\frac{\partial u}{\partial x} = \lim_{\delta x\to 0} \frac{f(x+\delta x, y) - f(x,y)}{\delta x}$$

Similarly, when y varies and x remains constant, the derivative of u with respect to y is known as the partial derivative of u with respect to y and it is denoted by $\partial u/\partial y$ or u_y and is defined as:

$$\frac{\partial u}{\partial y} = \lim_{\delta y\to 0} \frac{f(x, y+\delta y) - f(x,y)}{\delta y}$$

The partial derivative of $u = f(x, y)$ with respect to x at the point (x_0, y_0) is $\left.\frac{\partial u}{\partial x}\right|_{(x_0,y_0)} = \lim\limits_{\delta x\to 0} \frac{f(x_0+\delta x, y_0) - f(x_0,y_0)}{\delta x}$ and with respect to y at the point (x_0, y_0) is

$$\left.\frac{\partial u}{\partial y}\right|_{(x_0,y_0)} = \lim_{\delta y\to 0} \frac{f(x_0, y_0+\delta x) - f(x_0,y_0)}{\delta y} \text{ provided limits exist.}$$

Note: The partial derivative gives the rate of change of one variable at a point when other variables are kept constant.

EXAMPLE 6.9 Find the values of u_x and u_y at (1, 2) if $u = f(x, y) = 2x^2y - 3y^3 + 7$.

Ans: We have $u_x = \dfrac{\partial f}{\partial y} = 4xy$. So $u_x \Big|_{(1,2)} = 8$

$$u_y = \frac{\partial f}{\partial y} = 2x^2 - 9y^2. \text{ So } u_y \Big|_{(1,2)} = 2 - 36 = -34$$

EXAMPLE 6.10 Find $\dfrac{\partial f}{\partial x}$ of a function $f(x, y) = x \cos xy$.

Ans: $\dfrac{\partial f}{\partial x} = \dfrac{\partial}{\partial x}(x \cos xy) = \cos xy + xy \sin xy$.

EXAMPLE 6.11 Find f_x and f_y for all $f(x,y) = \dfrac{ax}{x + \tan y}$ using quotient rule.

Ans: $\dfrac{\partial f}{\partial x} = \dfrac{\partial}{\partial x}\left(\dfrac{ax}{x + \tan y}\right) = \dfrac{(x + \tan y)\dfrac{\partial}{\partial x}(ax) - ax\dfrac{\partial}{\partial x}(x + \tan y)}{(x + \tan y)^2}$

$$= \frac{a(x + \tan y) - ax}{(x + \tan y)^2} = \frac{a \tan y}{(x + \tan y)^2}$$

$$\frac{\partial f}{\partial y} = \frac{\partial}{\partial y}\left(\frac{ax}{x + \tan y}\right) = \frac{(x + \tan y)\dfrac{\partial}{\partial y}(ax) - ax\dfrac{\partial}{\partial y}(x + \tan y)}{(x + \tan y)^2}$$

$$= \frac{(x + \tan y)(0) - ax \sec^2 y}{(x + \tan y)^2} = \frac{-ax \sec^2 y}{(x + \tan y)^2}$$

EXAMPLE 6.12 Find $\partial u/\partial x$ if the function $xu - 1nu = x^2 + y^3$ defines u as a function of two variables x and y and the partial derivative exists.

Ans: Differentiate both sides of given expression with respect to x, keeping y constant.

$$\frac{\partial}{\partial x}(xu) - \frac{\partial}{\partial x}(\ln u) = \frac{\partial x^2}{\partial x} + \frac{\partial y^3}{\partial x}$$

$$x\frac{\partial u}{\partial x} + u - \frac{1}{u}\frac{\partial u}{\partial x} = 2x$$

$$\Rightarrow \left(x - \frac{1}{u}\right)\frac{\partial u}{\partial x} = 2x - u$$

$$\Rightarrow \left(\frac{ux-1}{u}\right)\frac{\partial u}{\partial x} = 2x - u$$

$$\Rightarrow \frac{\partial u}{\partial x} = \frac{2xu - u^2}{ux - 1}$$

EXAMPLE 6.13 The plane $x = 1$ intersects the paraboloid $u = x^2 + y^2$ in a parabola. Find the slope of the tangent to the parabola at (1, 2, 5).

Ans: The slope of tangent is the partial derivative $\partial u/\partial y$ evaluated at (1, 2).

$$\left.\frac{\partial u}{\partial y}\right|_{(1,\,2)} = \left.\frac{\partial}{\partial y}(x^2 + y^2)\right|_{(1,\,2)} = \left.2y\right|_{(1,\,2)} = 4$$

Function of More them Two Variables

EXAMPLE 6.14 Find $\partial u/\partial z$ for $u = y\cos(x + 3z) + z^2$.

Ans: $\dfrac{\partial u}{\partial z} = \dfrac{\partial}{\partial z}(y\cos(x + 3z) + z^2) = y\dfrac{\partial}{\partial z}\cos(x + 3z) + \dfrac{\partial}{\partial z}z^2 = 3y\sin(x + 3z) + 2z$

EXAMPLE 6.15 The resultant resistor of three resistors R_1, R_2, R_3 connected in parallel is given by $\dfrac{1}{R} = \dfrac{1}{R_1} + \dfrac{1}{R_2} + \dfrac{1}{R_3}$. Find the value of $\dfrac{\partial R}{\partial R_1}$ when $R_1 = 10$ ohms, $R_2 = 30$ ohms and $R_3 = 45$ ohms.

Ans: Here, we want to find $\dfrac{\partial R}{\partial R_1}$. So, treat R_2 and R_3 to be constant.

$$\frac{\partial}{\partial R_1}\left(\frac{1}{R}\right) = \frac{\partial}{\partial R_1}\left(\frac{1}{R_1} + \frac{1}{R_2} + \frac{1}{R_3}\right)$$

$$\frac{-1}{R^2}\frac{\partial R}{\partial R_1} = \frac{-1}{R^2}$$

$$\frac{\partial R}{\partial R_1} = \left(\frac{R}{R_1}\right)^2$$

Now $\dfrac{1}{R} = \dfrac{1}{10} + \dfrac{1}{30} + \dfrac{1}{45} = \dfrac{7}{45}$. Then $R = \dfrac{45}{7}$. So, $\dfrac{\partial R}{\partial R_1} = \left(\dfrac{45}{70}\right)^2 = \left(\dfrac{9}{14}\right)^2$

$= \dfrac{81}{196}$.

Second-Order Partial Derivatives

Let $u = f(x, y)$ be a function of two independent variables. In earlier section of this chapter, we have seen that $u_x = \frac{\partial f}{\partial x}$ and $u_y = \frac{\partial f}{\partial y}$. Differentiating these functions once again, we get $u_{xx} = \frac{\partial^2 f}{\partial x^2}$, $u_{xy} = \frac{\partial}{\partial y}\left(\frac{\partial f}{\partial x}\right) = \frac{\partial^2 f}{\partial y \partial x}$, $u_{yy} = \frac{\partial^2 f}{\partial y^2}$ and $u_{yx} = \frac{\partial}{\partial x}\left(\frac{\partial f}{\partial y}\right) = \frac{\partial^2 f}{\partial x \partial y}$. u_{xy} and u_{yx} are called mixed partial derivatives of f.

EXAMPLE 6.16 For $u = \tan^{-1}\frac{y}{x}$, prove that $\frac{\partial^2 u}{\partial x \partial y} = \frac{\partial^2 u}{\partial y \partial x}$.

Ans: Given $u = \tan^{-1}\left(\frac{y}{x}\right)$.

$$\therefore \qquad \frac{\partial u}{\partial x} = \frac{-y}{x^2+y^2} \text{ and } \frac{\partial u}{\partial y} = \frac{x}{x^2+y^2}$$

Then, $$\frac{\partial^2 u}{\partial y \partial x} = \frac{\partial}{\partial y}\left(\frac{\partial u}{\partial x}\right) = \frac{\partial}{\partial y}\left(\frac{-y}{x^2+y^2}\right) = -\frac{(x^2+y^2)1 - y(2y)}{(x^2+y^2)^2} = \frac{y^2-x^2}{(x^2+y^2)^2}$$

$$\frac{\partial^2 u}{\partial x \partial y} = \frac{\partial}{\partial x}\left(\frac{\partial u}{\partial y}\right) = \frac{\partial}{\partial x}\left(\frac{x}{x^2+y^2}\right) = \frac{(x^2+y^2)1 - x(2x)}{(x^2+y^2)^2} = \frac{y^2-x^2}{(x^2+y^2)^2}$$

Hence, $\frac{\partial^2 u}{\partial x \partial y} = \frac{\partial^2 u}{\partial y \partial x}$.

EXAMPLE 6.17 Find $u_x, u_y, u_{xy}, u_{xx}, u_{yy}, u_{yx}$ for the function $u(x, y) = y \cos x + xe^y$.

Ans: The first-order partial derivatives are

$$u_x = -y\sin x + e^y \text{ and } u_y = \cos x + xe^y$$

$$u_{xx} = \frac{\partial u_x}{\partial x} = \frac{\partial}{\partial x}(-y \sin x + e^y) = -y\cos x$$

$$u_{xy} = \frac{\partial u_x}{\partial y} = \frac{\partial}{\partial y}(-y \sin x + e^y) = -\sin x + e^y$$

$$u_{yy} = \frac{\partial u_y}{\partial y} = \frac{\partial}{\partial y}(\cos x + xe^y) = xe^y$$

$$u_{yx} = \frac{\partial u_y}{\partial x} = \frac{\partial}{\partial x}(\cos x + xe^y) = -\sin x + e^y$$

THEOREM 6.1 The Mixed Derivation Theorem
If $u = f(x, y)$ and its second-order partial derivatives u_x, u_y, u_{xy} and u_{yx} are well-defined in the neighborhood of a point (a, b) and are all continuous at (a, b) then $f_{xy}(a, b) = f_{yx}(a, b)$.

Differentiability

If $y = f(x)$ is differentiable at $x = x_0$, then the increment (or change) in the value of y when x_0 changes to $x_0 + \Delta x$ is given by $\Delta y = f'(x_0)\Delta x + \varepsilon \Delta x$ in which $\varepsilon \to 0$ as $\Delta x \to 0$.

Now, let us define increment in function of two variables.

THEOREM 6.2 Increment Theorem for Function of Two Variables
Assume that the first-order partial derivatives of $u = f(x, y)$ are defined in the neighborhood of (x_0, y_0) and f_x and f_y are continuous at (x_0, y_0). Then the change

$$\Delta u = f(x_0 + \Delta x, y_0 + \Delta y) - f(x_0, y_0)$$

in the value of u then (x_0, y_0) shifts to $(x_0 + \Delta x, y_0 + \Delta y)$ in the neighborhood of R satisfies the equation

$$\Delta u = f_x(x_0, y_0)\Delta x + f_y(x_0, y_0)\Delta y + \varepsilon_1 \Delta x + \varepsilon_2 \Delta y$$

in which $\varepsilon_1 \to 0$ as $\Delta x \to 0$ and $\varepsilon_2 \to 0$ as $\Delta y \to 0$.

Definition: A function $u = f(x, y)$ is differentiable at (x_0, y_0) if $f_x(x_0, y_0)$ and $f_y(x_0, y_0)$ exists and $\Delta u = f_x(x_0, y_0)\Delta x + f_y(x_0, y_0)\Delta y + \varepsilon_1 \Delta x + \varepsilon_2 \Delta y$ in which $\varepsilon_1 \to 0$ as $\Delta x \to 0$ and $\varepsilon_2 \to 0$ as $\Delta y \to 0$ is satisfied.

Note: A differentiable function $f(x, y)$ is at (x_0, y_0) is continuous at (x_0, y_0).

6.4 THE CHAIN RULE

We see here chain rule for function of one independent variable and two intermediate variables.

THEOREM 6.3 Let $u = f(x, y)$ be differentiable and $x = x(t)$ and $y = y(t)$ are differentiable functions of t. Then

$$\frac{du}{dt} = \frac{\partial u}{\partial x} \cdot \frac{dx}{dt} + \frac{\partial u}{\partial y} \cdot \frac{dy}{dt}$$

***Proof*:** Let Δx, Δy and Δu be the changes observed when t changes from t_0 to $t_0 + \Delta t$. Given u is differentiable.

Then $\Delta u = \left(\frac{\partial u}{\partial x}\right)_P \Delta x + \left(\frac{\partial u}{\partial y}\right)_P \Delta y + \varepsilon_1 \Delta x + \varepsilon_2 \Delta y$, where $P(x(t_0), y(t_0))$,

$\varepsilon_1 \to 0$ as $\Delta x \to 0$ and $\varepsilon_2 \to 0$ as $\Delta y \to 0$.

$$\frac{\Delta u}{\Delta t} = \left(\frac{\partial u}{\partial x}\right)_P \frac{\Delta x}{\Delta t} + \left(\frac{\partial u}{\partial y}\right)_P \frac{\Delta y}{\Delta t} + \varepsilon_1 \frac{\Delta x}{\Delta t} + \varepsilon_2 \frac{\Delta y}{\Delta t}$$

Taking limit $\Delta t \to 0$ gives $\dfrac{du}{dt} = \left(\dfrac{\partial u}{\partial x}\right)_p \dfrac{dx}{dt} + \left(\dfrac{\partial u}{\partial y}\right)_p \dfrac{dy}{dt}$.

EXAMPLE 6.18 Find the derivative of $u = xy$ with respect to t where, $x = \cos t$ and $y = \sin t$. Find value of the derivative at $t = \pi/4$.

Ans: $\dfrac{du}{dt} = \dfrac{\partial u}{\partial x}\dfrac{dx}{dt} + \dfrac{\partial u}{\partial y}\dfrac{dy}{dt} = -y\sin t + x\cos t = -\sin^2 + \cos^2 t = \cos 2t.$

Now, $\dfrac{du}{dt}$ at $t = \dfrac{\pi}{4}$ is $\cos\dfrac{\pi}{2} = 0$.

Chain rule for function of one independent variable t, three intermediate variables $x = x(t)$, $y = y(t)$ and $z = z(t)$ and dependent variable $u = f(x, y, z)$ is

$$\frac{du}{dt} = \frac{\partial u}{\partial x}\frac{dx}{dt} + \frac{\partial u}{\partial y}\frac{dy}{dt} + \frac{\partial u}{\partial z}\frac{dz}{dt}$$

EXAMPLE 6.19 Find du/dt it $u = yz + x, x = t\ y = \cos t, z = \sin t$

Ans: Using chain rule defined above,

$$\frac{du}{dt} = \frac{\partial u}{\partial x}\frac{dx}{dt} + \frac{\partial u}{\partial y}\frac{dy}{dt} + \frac{\partial u}{\partial z}\frac{dz}{dt} = 1 + z(-\sin t) + y\cos t = 1 - \sin^2 t + \cos^2 t = 2\cos^2 t$$

Chain Rule for Two Independent Variables and Three Intermediate Variables

Let $u = f(x, y, z)$, $x = g(r, s)$, $y = h(r, s)$ and $z = k(r, s)$ and be differentiable functions.

Then,
$$\frac{\partial u}{\partial r} = \frac{\partial u}{\partial x}\cdot\frac{\partial x}{\partial r} + \frac{\partial u}{\partial y}\cdot\frac{\partial y}{\partial r} + \frac{\partial u}{\partial z}\cdot\frac{\partial z}{\partial r}$$

$$\frac{\partial u}{\partial s} = \frac{\partial u}{\partial x}\cdot\frac{\partial x}{\partial s} + \frac{\partial u}{\partial y}\cdot\frac{\partial y}{\partial s} + \frac{\partial u}{\partial z}\cdot\frac{\partial z}{\partial s}$$

Let us illustrate this in following examples.

EXAMPLE 6.20 For $u = x^2 + 2y + z$, $x = 2r, y = r^2 + \ln s$ and $z = \dfrac{r}{s}$, find $\dfrac{\partial w}{\partial r}$ and $\dfrac{\partial w}{\partial s}$.

Ans: Given $u = x^2 + 2y + z$. Then

$$\frac{\partial u}{\partial x} = 2x, \frac{\partial u}{\partial y} = 2 \text{ and } \frac{\partial u}{\partial z} = 1$$

Now, for $x = 2r, \frac{\partial x}{\partial r} = 2$ and $\frac{\partial x}{\partial s} = 0.$

For $y = r^2 + \ln s, \frac{\partial y}{\partial r} = 2r$ and $\frac{\partial y}{\partial s} = \frac{1}{s}$

For $z = \frac{r}{s}, \frac{\partial z}{\partial r} = \frac{1}{s}$ and $\frac{\partial z}{\partial s} = -\frac{r}{s^2}$

Hence, $\frac{\partial w}{\partial r} = \frac{\partial u}{\partial x}.\frac{\partial x}{\partial r} + \frac{\partial u}{\partial y}.\frac{\partial y}{\partial r} + \frac{\partial u}{\partial z}.\frac{\partial z}{\partial r} = 2x(2) + 2(2r) + \frac{1}{s} = 8r + 4r + \frac{1}{s}$

$= 12r + \frac{1}{s}$

and $\frac{\partial w}{\partial s} = \frac{\partial u}{\partial x}.\frac{\partial x}{\partial s} + \frac{\partial u}{\partial y}.\frac{\partial y}{\partial s} + \frac{\partial u}{\partial z}.\frac{\partial z}{\partial s} = 2x(0) + 2\left(\frac{1}{s}\right) + 1\left(-\frac{r}{s^2}\right) = \frac{2}{s} - \frac{r}{s^2}$

If $u = f(x, y), x = g(r, s)$ and $y = h(r, s)$, then

$$\frac{\partial u}{\partial r} = \frac{\partial u}{\partial x}.\frac{\partial x}{\partial r} + \frac{\partial u}{\partial y}.\frac{\partial y}{\partial r} \text{ and } \frac{\partial u}{\partial s} = \frac{\partial u}{\partial x}.\frac{\partial x}{\partial s} + \frac{\partial u}{\partial y}.\frac{\partial y}{\partial s}$$

EXAMPLE 6.21 For $u = x^2 + y^2, x = r - s$ and $y = r + s$, find $\frac{\partial u}{\partial r}$ and $\frac{\partial u}{\partial s}$.

Ans: $u = x^2 + y^2$ then $\frac{\partial u}{\partial x} = 2x$ and $\frac{\partial u}{\partial y} = 2y.$

$x = r - s$ then $\frac{\partial x}{\partial r} = 1, \frac{\partial x}{\partial s} = -1$

$y = r + s$ then $\frac{\partial y}{\partial r} = 1, \frac{\partial y}{\partial s} = 1.$

So, $\frac{\partial u}{\partial r} = \frac{\partial u}{\partial x}.\frac{\partial x}{\partial r} + \frac{\partial u}{\partial y}.\frac{\partial y}{\partial r} = 2x(1) + 2y(1) = 2x + 2y = 4r$

and $\frac{\partial u}{\partial s} = \frac{\partial u}{\partial x}.\frac{\partial x}{\partial s} + \frac{\partial u}{\partial y}.\frac{\partial y}{\partial s} = 2x(-1) + 2y = 2(y - x) = 4s$

If and, then $u = f(x)$ and $x = g(p, s)$, then $\frac{\partial u}{\partial r} = \frac{du}{dx}.\frac{\partial x}{\partial r}$ and $\frac{\partial u}{\partial s} = \frac{du}{dx}.\frac{\partial x}{\partial s}$

Differentiation of Implicit Function

Let $f(x, y)$ be differentiable and the equation $f(x, y) = 0$ defines y as a function of x. Then at any point $\frac{dy}{dx} = -\frac{f_x}{f_y}$, provided $f_y \neq 0$.

$$u = f(x,y) = 0 \Rightarrow \frac{du}{dx} = 0.$$

By chain rule, derivation of $f(x, y)$ is $\frac{du}{dx} = f_x \frac{dx}{dx} + f_y \frac{dy}{dx} = 0.$

This gives $\frac{dy}{dx} = -\frac{f_x}{f_y}$ provided $f_y \neq 0$.

EXAMPLE 6.22 Find $\frac{dy}{dx}$ if $f(x,y) = ax^2 + 2hxy + by^2 + 2gx + 2fy + c = 0$.

Ans: We have $f_x = 2ax + 2hy + 2g$ and $f_y = 2hx + 2by + 2f$.

Then $$\frac{dy}{dx} = -\frac{fx}{fy} = -\frac{(2ax + 2hy + 2g)}{2hx + 2by + 2f} = -\left(\frac{ax + by + g}{hx + by + t}\right).$$

EXAMPLE 6.23 Find $\frac{dy}{dx}$ if $f(x,y) = x^y + y^x - a^b = 0$.

Ans: We have $f_x = yx^{y-1} + y^x \ln y$ and $f_y = x^y \ln x + xy^{x-1}$

So, $$\frac{dy}{dx} = -\frac{f_x}{f_y} = -\left(\frac{yx^{y-1} + y^x \ln y}{x^y \ln x + xy^{n-1}}\right).$$

Given $u = f(x, y)$, Differentiating $f(x, y, u) = 0$, partially with respect to x and y respectively gives

$$\frac{\partial F}{\partial x} + \frac{\partial F}{\partial u}.\frac{\partial u}{\partial x} = 0 \Rightarrow \frac{\partial u}{\partial x} = -\frac{F_x}{F_u}$$

and $\frac{\partial F}{\partial y} + \frac{\partial F}{\partial u}.\frac{\partial u}{\partial y} = 0 \Rightarrow \frac{\partial u}{\partial y} = -\frac{F_y}{F_u}$ provided $F_u \neq 0$.

Note: This result is known as **Implicit Function Theorem**.

EXAMPLE 6.24 If three thermodynamic variables P, V, T satisfies a relation $F(P, V, T) = 0$, show that $\left(\frac{\partial P}{\partial T}\right)_V \left(\frac{\partial T}{\partial V}\right)_P \left(\frac{\partial V}{\partial P}\right)_T = -1.$

Ans: Taking differentials of $F(P, V, T) = 0$, we get $\frac{\partial F}{\partial P} dP + \frac{\partial F}{\partial V} dV + \frac{\partial F}{\partial T} = 0$

It V is constant then $dV = 0$ and $\dfrac{\partial F}{\partial P}\dfrac{dP}{dT} + \dfrac{\partial F}{\partial T} = 0$

Since V is constant, one can write $\dfrac{dP}{dT} = \left(\dfrac{\partial T}{\partial T}\right)_V$

$$\left(\frac{\partial P}{\partial T}\right)_V = -\frac{F_T}{F_P}$$

Similarly, $\left(\dfrac{\partial T}{\partial V}\right)_P = -\dfrac{F_V}{F_T}$ and $\left(\dfrac{\partial V}{\partial P}\right)_T = -\dfrac{F_P}{F_V}$

Multiplying these three gives $\left(\dfrac{\partial P}{\partial T}\right)_V \left(\dfrac{\partial T}{\partial V}\right)_P \left(\dfrac{\partial V}{\partial P}\right)_T = -1.$

6.5 DIRECTIONAL DERIVATIVES AND GRADIENT VECTORS

It has been seen earlier in this chapter that the rate of change of function defines direction or slope of the tangent to a curve at a point.

Directional Derivatives in a Plane

If $f(x, y)$ with $x = g(t)$, $y = h(t)$ are differentiable functions, then the rate at which f changes with respect to t is given by $\dfrac{df}{dt} = \dfrac{\partial f}{\partial x}\dfrac{dx}{dt} + \dfrac{\partial f}{\partial y}\dfrac{dy}{dt}$.

At any point $P(x_0, y_0)$, this gives the rate of change of f with respect to increasing t and three for depends on the motion along the curve. If a curve is a straight line and t is the arc length parameter along the line paring through P in the direction of a given unit vector $\hat{n}$, the $\partial f/\partial t$ is the rate of change of f with respect to distance in its domain in the direction of $\hat{n}$. One can find the rate at which f changes with respect to distance at it passes through P in different directions by varying $\hat{n}$.

Let $f(x, y)$ be a function defined in the region R in the xy-plane and $P(x_0, y_0) \in R$. Let $\hat{n} = n_1\hat{i} + n_2\hat{j}$ be a unit Vector. Then the equations $x = x_0 + tn_1$, $y = y_0 + tn_2$.

Parameterized the line through P parallel to $\hat{n}$.

Definition: The derivative of f at $P(x_0, y_0)$ in the direction of the unit vector $\hat{n} = n_1\hat{i} + n_2\hat{j}$ is

$$\left(\frac{\partial f}{\partial t}\right)_{\hat{n}P} = \lim_{t \to 0} \frac{f(x_0 + tn_1, y_0 + tn_2) - f(x_0, y_0)}{t}$$

provided limit exists. We denote it by $(D_{\hat{n}}t)_P$ and read it as the derivative of f at P in the direction of $\hat{n}$. The partial derivatives $f_x(x_0, y_0)$ and $f_y(x_0, y_0)$ are the directional derivatives of f at P in the $\hat{i}$ and $\hat{j}$ directions.

EXAMPLE 6.25 Find the derivatives of $f(x,y)=y^2+2xy$ at $P(1, 1)$ in the direction of the unit vector $\hat{n}=(1/2)\hat{i}+(\sqrt{3}/2)\hat{j}$.

Ans: Applying equation written above at $P(1,1)$, we get

$$\left(\frac{\partial f}{\partial t}\right)=\lim_{t\to 0}\frac{f(x_0+tn_1,y_0+tn_2)-f(x_0,y_0)}{t}=\lim_{t\to 0}\frac{f\left(1+t\frac{1}{2},1+t\frac{\sqrt{3}}{2}\right)-f(1,1)}{t}$$

$$=\lim_{t\to 0}\frac{1}{t}\left(\left(\frac{3}{4}+\frac{\sqrt{3}}{2}\right)t^2+2\sqrt{3}t+t\right)=\lim_{t\to 0}\left(\left(\frac{3}{4}+\frac{\sqrt{3}}{2}\right)t+2\sqrt{3}+1\right)=1+2\sqrt{3}$$

Therefore, the rate of change of $f(x,y)=y^2+2xy$ at $P(1, 1)$ in the direction of the unit vector $\hat{n}=(1/2)\hat{i}+(\sqrt{3}/2)\hat{j}$ is $1+2\sqrt{3}$.

Let us understand physical interpretation of the directional derivative.

Suppose $T(x, y)$ is the temperature at every point (x, y) over *a* region in the *xy*-plane. Then $T(x_0, y_0)$ is the temperature at a point $P(x_0, y_0)$ and $(D_{\hat{n}} f)_P$ governs the instantaneous rate of change of temperature at P in the direction $\hat{n}$.

Gradient of *f*(*x*, *y*)

Let $x = x_0 + tn_1$ and $y = y_0 + tn_2$. We have

$$\left(\frac{df}{dt}\right)_{\hat{n}P}=\left(\frac{\partial f}{\partial x}\right)_P\frac{dx}{dt}+\left(\frac{\partial t}{\partial y}\right)_P\frac{\partial y}{\partial t}=\left(\frac{\partial f}{\partial x}\right)_P n_1+\left(\frac{\partial f}{\partial y}\right)_P n_2$$

$$=\left[\left(\frac{\partial f}{\partial x}\right)_P\hat{i}+\left(\frac{\partial f}{\partial y}\right)_P\hat{j}\right]\cdot(n_1\hat{i}+n_2\hat{j})$$

This derivation leads to the definition of gradient of function at a point.

Definition: The gradient vector of $f(x, y)$ at a point $P(x_0, y_0)$ is the vector $\nabla f=\frac{\partial f}{\partial x}\hat{i}+\frac{\partial f}{\partial y}\hat{j}$ obtained by computing the partial derivatives of f at $P(x_0, y_0)$.

Note: The notation ∇f is read as "grad f".

THEOREM 6.4 The Directional Derivative is a Dot product.

If $f(x, y)$ is differentiable in the neighborhood of $P(x_0, y_0)$, then $\left(\frac{\partial f}{\partial t}\right)_{\hat{n}P}=(\nabla f)_P\cdot\hat{n}$ the dot product of the gradient ∇f at P and $\hat{n}$.

EXAMPLE 6.26 Find the derivative of $f(x, y) = ye^x + \sin(xy)$ at point (0, 1) in the direction of $\vec{n}=3\hat{i}-4\hat{j}$.

Ans: The unit vector $\hat{n} = \frac{\vec{n}}{|n|} = \frac{3}{5}\hat{i} - \frac{4}{5}\hat{j}$. The partial derivatives of $f(x, y)$are

$$f_x(x,y) = ye^x + y\cos(xy) \qquad f_y(x,y) = e^x + x\cos(xy).$$

Then $\qquad f_x(0, 1) = 2 \qquad\qquad f_y(0, 1) = 1$

The gradient of f at (0, 1) is $\nabla f_{(0,\,1)} = f_x(0,1)\hat{i} + f_y(0,1)\hat{j} = 2\hat{i} + \hat{j}$.

The derivative of f at (0, 1) in the direction of $\vec{n}$ is

$$(D\hat{n}f)_p = \nabla f_{(0,\,1)} \cdot \hat{n} = (2\hat{i} + \hat{j}) \cdot \left(\frac{3}{5}\hat{i} - \frac{4}{5}\hat{j}\right) = \frac{2}{5}$$

The Dot Product Formula

$$(D_{\hat{n}}f)_P = \nabla f \cdot n = |\nabla f||n|\cos\theta$$

where θ is the angle between ∇f and vector $\vec{n}$. This definition characterized ∇f as follows:

1. The function f increases faster. When $\cos\theta = 1$ or when $\cos\theta = 0$ and $\vec{n}$ is the direction of ∇f. One can observe that the derivative in this direction is $D_{\vec{n}}f = |\nabla f|\cos 0 = |\nabla f|$.
2. f decreases most rapidly in the direction of $-\nabla f$. The derivative in this direction is $D_{\vec{n}}f = |\nabla f|\cos\pi = -|\nabla f|$.
3. Gradient of f is orthogonal to $\vec{n}$ if $D_{\vec{n}}f = 0$.

Let us illustrate these properties in next example.

EXAMPLE 6.27 Find the direction in which $f(x, y) = 4x^2 + 9y^2$.

(a) increase most rapidly at the point (1, 1).
(b) decrease most rapidly at (1, 1).
(c) What are the direction of zero change in f at (1, 1)?

Ans: (a) The gradient of function at $\left(\frac{1}{2}, \frac{1}{3}\right)$

$$\vec{u} = (\nabla f) = 8x\hat{i} + 9y\hat{j} = \hat{i} + 3\hat{j} \text{ at } \left(\frac{1}{2}, \frac{1}{3}\right)$$

Its direction is $\hat{u} = \frac{\vec{u}}{|\vec{u}|} = \frac{4}{5}\hat{i} + \frac{3}{5}\hat{j}$

The function increases most rapidly in the direction of $\frac{4}{5}\hat{i} + \frac{3}{5}\hat{j}$.

(b) The function decreases most rapidly in the direction of $-\frac{4}{5}\hat{i}-\frac{3}{5}\hat{j}$.

(c) The direction of zero change at $\left(\frac{1}{2},\frac{1}{3}\right)$ is the direction orthogonal to $\nabla f:\vec{n}=-\frac{4}{5}\hat{i}+\frac{3}{5}\hat{j}$ and $-\vec{n}=\frac{4}{5}\hat{i}-\frac{3}{5}\hat{j}$.

Gradients and Tangents to Level Curves

If a differentiable function $f(x, y)$ takes a constant value c along a smooth curve $\vec{u}=g(t)\hat{i}+h(t)\hat{j}$, then $f(g(t), h(t))=c$.

Differentiating both sides with respect to t gives

$$\frac{\partial f}{\partial x}\frac{dg}{dt}+\frac{\partial f}{\partial y}\frac{dh}{dt}=0$$

$$\Rightarrow\left(\frac{\partial f}{\partial x}\hat{i}+\frac{\partial f}{\partial y}\hat{j}\right)\cdot\left(\frac{dg}{dt}\hat{i}+\frac{dh}{dt}\hat{j}\right)=0$$

This suggests that ∇f is normal to the tangent vector $\frac{d\vec{u}}{dt}$, so it is normal to the curve. $\nabla f\cdot\frac{d\vec{u}}{dt}=0$ defines perpendicular to the level curves. This helps us to find equations for tangent lines to level curves. These are the lines normal to the gradients. The line through a point $P(x_0, y_0)$ to a vector $\vec{n}=a\hat{i}+b\hat{j}$ has the equation $a(x-x_0)+b(y-y_0)=0$.

If $\vec{n}$ is the gradient $(\nabla f)_P=f_x(x_0,y_0)\hat{i}+f_y(x_0,y_0)\hat{j}$, the equation of tangent line is $f_x(x_0,y_0)(x-x_0)+f_y(x_0,y_0)(y-y_0)=0$.

EXAMPLE 6.28 Find an equation for the tangent to hyperbola $\frac{x^2}{4}=\frac{y^2}{9}=1$ at the point (2, 0).

Ans: The hyperbola is a level curve of the function $f(x,y)=\frac{x^2}{4}-\frac{y^2}{9}-1=0$.

$$\nabla f=\frac{x}{2}\hat{i}-\frac{2y}{9}\hat{j}=1.\hat{i}+0.\hat{j}$$

The tangent is the line $1(x-2)+0(y-0)=0\Rightarrow x=2$.

Algebraic Properties for Gradients

1. Sum Rule: $\nabla(f+g)=\nabla f+\nabla g$
2. Difference Rule: $\nabla(f-g)=\nabla f-\nabla g$

3. Scalar Multiplication Rule: $\nabla(cf) = c\nabla f$, c is any scalar
4. Product Rule: $\nabla(fg) = f\nabla g + g\nabla f$
5. Quotient Rule: $\nabla\left(\dfrac{f}{g}\right) = \dfrac{g\nabla f - f\nabla g}{g^2}$

Function of Three Variables

For a differentiable function $f(x, y, z)$ and a unit vector $\vec{n} = n_1\hat{i} + n_2\hat{j} + n_3\hat{k}$ in space,

$$\nabla f = \frac{\partial f}{\partial x}\hat{i} + \frac{\partial f}{\partial y}\hat{j} + \frac{\partial f}{\partial z}\hat{k}$$

and

$$D_n f = \nabla f \cdot \hat{n} = \frac{\partial f}{\partial x}n_1 + \frac{\partial f}{\partial y}n_2 + \frac{\partial f}{\partial z}n_3.$$

EXAMPLE 6.29 Find the derivative of $f(x, y, z) = yx^2 + y^3 + z$ at (1,1,0) in the direction of $\vec{n} = \hat{i} - \hat{j} + \hat{k}$.

In what direction does f change most rapidly at (1,1,0) and what are the rates of change in these directions?

Ans: Given $\vec{n} = \hat{i} - \hat{j} + \hat{k}$. Then unit vector $\hat{n} = \dfrac{\vec{n}}{|n|} = \dfrac{\hat{i}}{\sqrt{3}} - \dfrac{\hat{j}}{\sqrt{3}} + \dfrac{k}{\sqrt{3}}$.

Now, $\quad f_x = 2xy,\ f_y = x^2 + 3y^2,\ f_z = 1$

At $\quad (-1, 1, 0),\ f_x = -2,\ f_y = 4,\ f_z = 1$

$\therefore \quad \nabla f = -2\hat{i} + 4\hat{j} + \hat{k}$

The derivative of f at (–1,1,0) in the direction of $\vec{n}$ is

$$(D_{\vec{n}} f)_{(-1,1,0)} = \nabla f_{(-1,1,0)} \cdot \hat{n} = (-2\hat{i} + 4\hat{j} + \hat{k}) \cdot \left(\frac{1}{\sqrt{3}}\hat{i} - \frac{1}{\sqrt{3}}\hat{j} + \frac{1}{\sqrt{3}}k\right)$$

$$= -\frac{2}{\sqrt{3}} + \frac{4}{\sqrt{3}} + \frac{1}{\sqrt{3}} = \sqrt{3}$$

The function increases most rapidly in the direction $\nabla f = -2\hat{i} + 4\hat{j} + \hat{k}$ and decreases most rapidly in the direction of $-\nabla f$. The rates of change in the directions are $|\nabla f| = \sqrt{21}$ and $-|\nabla f| = -\sqrt{21}$.

6.6 TANGENT PLANES AND DIFFERENTIALS

In this section, we will discuss about tangent plane at a point on a smooth surface in space.

Tangent Plane and Normal Lines

Let $f(x, y, z) = c$ be a differentiable function and $\vec{n} = g(t)\hat{i} + h(t)\hat{j} + k(t)\hat{k}$ be a smooth curve on the level surface $f(x, y, z) = c$.

Differentiating, $f(g(t), h(t), k(t)) = c$ gives

$$\frac{\partial f}{\partial x}\frac{dg}{dt} + \frac{\partial f}{\partial y}\frac{dh}{dt} + \frac{\partial f}{\partial z}\frac{dk}{dt} = 0 \Rightarrow \left(\frac{\partial f}{\partial x}\hat{i} + \frac{\partial f}{\partial y}\hat{j} + \frac{\partial f}{\partial z}\hat{k}\right)\cdot\left(\frac{dg}{dt}\hat{i} + \frac{dh}{dt}\hat{j} + \frac{dk}{dt}\hat{k}\right) = 0$$

At every point along the curve, ∇f is normal to the velocity vector of the curve.

Definition: The tangent plane at the point $P(x_0, y_0, z_0)$ on the level surface $f(x, y, z) = c$ of a differentiable function f is the plane through P normal to ∇f at P.

The normal line of the surface at P is the line through P parallel to ∇f at P.

The equation of the tangent plane at P is $f_x(P)(x - x_0) + f_y(P)(y - y_0) + f_z(P)(z - z_0) = 0$ and that of normal line is $x = x_0 + f_x(P)t$, $y = y_0 + f_y(P)t$, $z = z_0 + f_z(P)t$.

EXAMPLE 6.30 Find the tangent plane and normal line of the surface $f(x, y, z) = x^2 + y^2 + 2z - 6 = 0$ at the point $P(2, 2, -1)$.

Ans: We have $f_x = 2x$, $f_y = 2y$ and $f_z = 2$ then gradient at P is $4\hat{i} + 4\hat{j} - 2\hat{k}$. The equation of tangent plane is

$$4(x-2) + 4(y-2) - 2(z+1) = 0 \Rightarrow 4x + 4y - 2z + 14 = 0 \Rightarrow 2x + 2y - z + 7 = 0$$

The normal line at P is $x = 2 + 4t$, $y = 2 + 4t$, $z = -1 - 2t$.

The equation for the plane tangent to a smooth surface $z = f(x, y)$ at a point $P(x_0, y_0, z_0)$ with $z_0 = f(x_0, y_0)$. Define $F(x, y, z) = f(x - y) - z$.

Then $f_x(P)(x - x_0) + f_y(P)(y - y_0) - (z - z_0) = 0$ is equation for the plane tangent to the level surface at P because $F_x = f_x$, $F_y = f_y$ and $F_z = -1$.

EXAMPLE 6.31 Find the plane tangent to the surface $z = xe^y - y\cos x$ at $(0, 0, 0)$.

Ans: Let $f(x, y) = xe^y - y\cos x$. Then

$$f_x(x, y) = e^y + y\sin x \text{ and } f_y(x, y) = xe^y - \cos x$$

At $(0, 0)$, $f_x(0, 0) = 1$ and $f_y(0, 0) = -1$

The equation of tangent plane is

$$1(x - 0) - 1(y - 0) - (z - 0) = 0 \Rightarrow x - y - z = 0$$

EXAMPLE 6.32 Find parametric equation for the line tangent to Q at the point $P(1, -1, 2)$ where Q is the point of intersection of the surface $f(x, y, z) = x^2 + y^2 - 3 = 0$ and $g(x, y, z) = x + z - 6 = 0$.

Ans: The tangent is orthogonal to both ∇f and ∇g at P and therefore parallel to

$$\vec{n} = \nabla f \times \nabla g. \text{ At } P(1, -1, 2),$$

$$\nabla f\Big|_P = (2x\hat{i} + 2y\hat{j})_P = 2\hat{i} - 2\hat{j} \text{ and } \nabla g\Big|_P = (\hat{i} + \hat{k})_P = \hat{i} + \hat{k}$$

$$\vec{n} = \nabla f \times \nabla g = \begin{vmatrix} \hat{i} & \hat{j} & \hat{k} \\ 2 & -2 & 0 \\ 1 & 0 & 1 \end{vmatrix} = -2\hat{i} - 2\hat{j} + 2\hat{k}$$

The parametric equation for the tangent line are $x = 1 - 2t$, $y = -1 - 2t$ and $z = 1 + 2t$.

Computing Change in a Specific Direction

To find the change in the value of a function f, when it is moved by a small distance dh from a point P.

For a single variable function, it is $df = f'(P)dh$ and for a function of two or more variables, it is given by $df = \left(\nabla f\big|_P \cdot \vec{n}\right) dh$, where $\vec{n}$ is the direction of the motion away from P.

EXAMPLE 6.33 Compute the change in the value of the function $f(x, y, z) = x \cos y + 2x^2 + yz$ if the point moves 0.1 unit from $P(1,0,0)$ towards $Q(2,1,-1)$.

Ans: The direction of the vector $\overrightarrow{PQ} = \vec{n} = \hat{i} + \hat{j} + \hat{k}$

$$\vec{n} = \frac{1}{\sqrt{3}}\hat{i} + \frac{1}{\sqrt{3}}\hat{j} + \frac{1}{\sqrt{3}}\hat{k}$$

The gradient of f at $P(1,0,0)$ is

$$\nabla f\Big|_P = (\cos y + 4x)\hat{i} + (-x \sin y + z)\hat{j} + \hat{k}\,\Big|_P = 5\hat{i}$$

Then, $\nabla f\big|_P \cdot \vec{n} = \dfrac{5}{\sqrt{3}}$

The change ∇f in f for $dh = 0.1$ away from P in the direction of $\hat{n}$ is

$$df = \left(\nabla f\Big|_P \cdot \vec{n}\right) dh = \frac{5}{\sqrt{3}}(0.1) = 0.2887 \text{ units.}$$

Definition: The linearization of a differentiable function $f(x, y)$ at (x_0, y_0) is $L(x,y) = f(x_0,y_0) + f_x(x_0,y_0)(x - x_0) + f_y(x_0,y_0)(y - y_0)$.

The estimate $f(x, y) = L(x, y)$ is called the standard linear approximation of f at (x_0, y_0).

EXAMPLE 6.34 Find the linearization of $f(x,y) = y^2 - xy + x^2 + 8$ at the point $P(1,2)$.

Ans: Let us calculate f, f_x and f_y at the point P.

$$f(1,2) = 11, f_x(1,2) = (-y+2x)_{(1,2)} = 0 \text{ and } f_y(1, 2) = (2y-x)_{(1,2)} = 3$$

Then linearization of f at (1,2) is

$$L(x, y) = f(1, 2) + f_x(1,2)(x-1) + f_y(1,2)(y-2) = 11 + 3(y-2) = 3y+5$$

The error E throughout the rectangular region $R : |x - x_0| \leq h,\ |y - y_0| \leq k$ can be computed if common upper bound M for $|f_{xx}|$, $|f_{xy}|$ and $|f_{yy}|$ exist at (x_0, y_0). The error is defined by $E(x, y) = f(x, y) - L(x, y)$.

The Error in the Standard Linear Approximation

Let f possesses continuous first and second order partial derivatives in an open set containing a rectangle R centered at (x_0, y_0). Let M be the upper bound for the values of $|f_{xx}|$, $|f_{yy}|$ and $|f_{xy}|$ on R. The error $E(x, y)$ observed in replacing $f(x, y)$ on R by its linearization $c(x, y) = f(x_0, y_0) + f_x(x_0, y_0)(x - x_0) + f_y(x_0, y_0)(y - y_0)$ satisfies the inequality $|E(x, y)| \leq \frac{1}{2} M(|x - x_0| + |y - y_0|)^2$.

EXAMPLE 6.35 Find the upper bound the error in the approximation $f(x, y)$ given in example 3.34 over the rectangle $R : |x-1| \leq 0.1,\ |y-2| \leq 0.1$. Express the upper bound as a percentage of $f(1,2)$, the value of f at the center of the rectangle.

Ans: We have $f_{xx} = 0$, $f_{yy} = 0$ and $f_{yy} = 2$. Define upper bound to be 2.

$$\therefore \qquad |E(x, y)| \leq \frac{1}{2} \cdot 2(|x-1| + |y-2|)^2$$

Given $|x-1| \leq 0.1$ and $|y-2| \leq 0.1$ on R. So, we have

$$|E(x, y)| \leq (0.1 + 0.1)^2 = 0.04$$

Now $f(1, 2) = 11$ so percentage error is $\dfrac{0.04}{11} \times 100 = 0.36\%$.

Differentials

For a single valued function $y = f(x)$ the change in f as x changes from a to $a + \Delta x$ is given by $\Delta f = f(a + \Delta x) - f(a)$ and the differential of f is given by $df = f'(a)\Delta x$.

Similarly, for a function $f(x, y)$ of two variables, the change in f is given by $\Delta f = f(x_0 + \Delta x, y_0 + \Delta y) - f(x_0, y_0)$ when a point (x_0, y_0) changes to $(x_0+\Delta x, y_0+\Delta y)$. Put $dx=\Delta x = x - x_0$ and $dy=\Delta y = y-y_0$. Then $df = f_x(x_0, y_0)dx + f_y(x_0, y_0)dy$ denotes the total differential of *f*.

6.7 EXTREME VALUES AND SADDLE POINTS

Let $f(x, y)$ be defined on a region R containing the point (a, b). Then

1. $f(a, b)$ is a local maximum value of function if $f(a,b)\geq f(x,y)$ for all points (x, y) in an open disk centered at (a, b).
2. $f(a, b)$ is a local minimum value of function *f* if $f(a,b)\leq f(x,y)$ for all points (x, y) in an open disk centered at (a, b).

Note: Local maxima correspond to mountain peaks on the surface $z = f(x, y)$ and local minima correspond to valley bottom.

Note: At these points the tangent planes, when they exist, are horizontal.

Note: Local extrema are also known as relative extrema.

THEOREM 6.5 First Derivative Test for Local Extreme Values
If $f(x, y)$ has a local maximum or minimum value at an interior point (a, b) and the first order partial derivatives exists at (a, b), then $f_x(a, b) = 0$ and $f_y(a, b) = 0$.

Proof: If *f* has a local extrema at (a, b), then the function $g(x) = f(x, b)$ has a local extrema at $x = a$. So, $g'(a) = 0$. Now $g'(a) = f_x(a, b)$, so $f_x(a, b) = 0$. Similarly, one can prove $f_y(a, b) = 0$.

Definition: An interior point of the domain of a function $f(x, y)$ where both f_x and f_y are zero or where are or both of f_x and f_y does not exist is called a critical point of *f*.

Definition: A differentiable function $f(x, y)$ has a saddle point at a critical point (a, b) if $f(x, y) > f(a, b)$ and $f(x, y) < f(a, b)$ for all (x, y) in every open disk centered of (a, b). The corresponding point $(a, b, f(a, b))$ on the surface $z = f(x, y)$ is called a saddle point of the surface.

EXAMPLE 6.36 Find the local extreme values of $f(x, y) = x^2 + y^2 - 2x - 4y + 10$.

Ans: We have $f_x = 2x - 2$ and $f_y = 2y - 4$. For local extreme values, $f_x = 2x - 2 = 0$ and $f_y = 2y - 4 = 0$. This gives $x = 1$ and $y = 2$. So, we have extreme point as (1, 2). At (1, 2), value of *f* is 6. Also, $x^2 + y^2 - 2x - 4y + 10 = (x - 1)^2 + (y - 2)^2 + 5$ which is never less than 5, guarantees that the critical point (1, 2) gives a local minimum.

THEOREM 6.6 Second Derivative Test for Local Extreme Values
Let $f(x, y)$ and its first and second order continuous partial derivative in a disk centered at (a, b) exists and $f_x(a, b) = f_y(a, b) = 0$ then

1. f has a local maxima of (a, b) if $f_{xx} < 0$ and $f_{xx}f_{yy} - f_{xy}^2 > 0$ at (a, b).
2. f has a local minima at (a, b) if $f_{xx} > 0$ and $f_{xx}f_{yy} - f_{xy}^2 > 0$ at (a, b).
3. f has a saddle point at (a, b) if $f_{xx}f_{yy} - f_{xy}^2 < 0$ at (a, b).
4. Test fails at (a, b) if $f_{xx}f_{yy} - f_{xy}^2 = 0$ at (a, b).

Note: $f_{xx}f_{yy} - f_{xy}^2$ is termed as the discriminant or Hessian of f. One can write $f_{xx}f_{yy} - f_{xy}^2$ as $\begin{vmatrix} f_{xx} & f_{xy} \\ f_{xy} & f_{yy} \end{vmatrix}$.

Geometrically, if the Hessian of f is positive at the point (a, b), then the curve shape is downward. if $f_{xx} < 0$, suggesting a local maximum and upward if $f_{xx} > 0$, suggesting a local minimum. On the other hand, if the Hessian of f is negative at a critical point (a, b) then the surface curves up in some directions and down in others resulting in to a saddle point.

EXAMPLE 6.37 Find the local extreme values of the function $f(x, y) = x^2 + y^2 - xy + 2x + 2y - 4$.

Ans: f is well-defined for all x and y. At extreme values, f_x and f_y are simultaneously zero.

So, $f_x = 2x - y + 2 = 0$ and $f_y = 2y - x + 2 = 0$ $\therefore$ $x = -2$ and $y = -2$. $(-2, -2)$ is an extreme point of f.

Now, $f_{xx} = 2$ $f_{xy} = -1$ $f_{yy} = 2$. Then $f_{xx}f_{yy} - f_{xy}^2 = 4 - 1 = 3 = 0$.

Thus, we have $f_{xx} = 2 > 0$ and $f_{xx}f_{yy} - f_{xy}^2 = 3 > 0$. Suggesting that $(-2, -2)$ is point of local minima at $(-2, -2)$. The local minimum value of f at this point is $f(-2, -2) = -8$.

6.8 LAGRANGE MULTIPLIERS

For constrained optimization problem, we use Lagrange multiplies.

EXAMPLE 6.38 Find the point $P(x, y, z)$ on the plane $x + 2y - z = 5$ that is closest to the origin.

Ans: We want to minimize distance between $O(0,0,0)$ and $P(x,y,z)$ subject to the constraint $x + 2y - z = 5$. Given $f(x, y, z) = OP^2 = x^2 + y^2 + z^2$ and $z = x + 2y - 5$. So, the problem is to find point (x, y) at which $g(x,y) = f(x, y, x + 2y - 5) = x^2 + y^2 + (x + 2y - 5)^2$ attains its minimum value. We have

$$g_x = 2x + 2(x + 2y - 5) = 0,\ g_y = 2y + 4(x + 2y - 5) = 0$$

This gives

$4x + 4y - 10 = 0$ or $2x + 2y - 5 = 0$ and $4x + 10y - 20 = 0$ or $2x + 5y - 10 = 0$

Solving this gives $x = \frac{5}{6}, y = \frac{5}{3}$ and hence $z = \frac{-5}{6}$. Hence, the closest point is $\left(\frac{5}{6}, \frac{5}{3}, \frac{-5}{6}\right)$.

The Method of Lagrange Multipliers

In this method, the extreme values of a function $f(x, y, z)$ subject to a constraint $g(x, y, z) = 0$ can be found among the point where $\nabla f = \lambda \nabla g$ for some scalar λ. This λ is called Lagrange multiplier.

THEOREM 6.7 The Orthogonal Gradient Theorem

Let $f(x, y, z)$ be a differentiable in a region whose interior contains a smooth curve $c : \vec{r}(t) = g(t)\hat{i} + h(t)\hat{j} + k(t)\hat{k}$. If P is a point on c where f has a local maximum or minimum relative to its values on c, then ∇f is orthogonal to c at P.

***Proof*:** The value of f at c is $f(g(t), h(t), k(t))$. The derivative with respect to t is $\frac{df}{dt} = \frac{\partial f}{\partial x}\frac{dg}{dt} + \frac{\partial f}{\partial y}\frac{dh}{dt} + \frac{dg}{\partial z}\cdot\frac{dk}{dt} = \nabla f \cdot n$ takes value zero at any point P where f has a local maximum or minimum. So, $\nabla f \cdot \vec{n} = 0$.

Note: At the points on a smooth curve $\vec{r}(t) = g(t)\hat{i} + h(t)\hat{j}$ where a differentiable function $f(x, y)$ is local maxima and minima relative to its values on the curve, $\nabla f \cdot \vec{n} = 0$ where $\vec{n} = d\vec{r}/dt$.

Let $f(x, y, z)$ and $g(x, y, z)$ be differentiable with $\nabla g \neq 0$ when $g(x, y, z) = 0$ to compute the local maximum and minimum values of f subject to the constraint $g(x, y, z) = 0$ (if exists), find the values of x, y, z and λ which satisfies $\nabla f = \lambda \nabla g$ and $g(x, y, z) = 0$ simultaneously.

EXAMPLE 6.39 Find the maxima and minimum values of the function $f(x, y) = xy$ takes on the parabola $4x^2 + 16y^2 = 64$.

Ans: We are asked to find the extreme values of $f(x, y) = xy$ subject to the constraint $g(x, y) = 4x^2 + 16y^2 - 64 = 0$. For this, we need to solve $\nabla f = \lambda \nabla g$ and $g(x, y) = 0$ simultaneously.

$$\nabla f = y\hat{i} + x\hat{j} \text{ and } \nabla g = 8x\hat{i} + 32y\hat{j} \Rightarrow y\hat{i} + x\hat{j} = 8\lambda x\hat{i} + 32\lambda y\hat{j}$$

$$\Rightarrow y = 8\lambda x \text{ and } x = 32\lambda y \Rightarrow x = 32\lambda y \text{ and } y = 256\lambda^2 y$$

So, $y = 0$ or $\lambda = \pm 16$. If $y = 0$ then $x = 0$. But (0,0) is not an the ellipse. Hence, $y \neq 0$. This suggests $\lambda = \pm\frac{1}{16}$ and $x = \pm 2y$. We substitute this in $4x^2 + 16y^2 = 64$. Then $y = \pm 2$ and hence $x = \pm 4$. Therefore the function f takes extreme values on the ellipse at the four points $(\pm 4, 2)$ and $(\pm 4, -2)$ as 8 or -8.

EXAMPLE 6.40 Find the maxima and minimum values of the function $f(x, y) = 2x + 3y$ on the circle $x^2 + y^2 = 4$.

Ans: We have $f(x,y) = 2x + 3y$, $g(x,y) = x^2 + y^2 - 4$. Find the values of x, y and λ that satisfies the equation $\nabla f = \lambda \nabla g \Rightarrow 2\hat{i} + 3\hat{j} = 2x\lambda\hat{i} + 2y\lambda\hat{j} \Rightarrow 2 = 2y\lambda$ and $3 = 2y\lambda$.

$\Rightarrow x = \frac{1}{\lambda}$ and $y = \frac{3}{2\lambda}$. This point $\left(\frac{1}{\lambda}, \frac{3}{2\lambda}\right)$ lies on the circle. So $\frac{1}{\lambda^2} + \frac{9}{4\lambda^2} = 4 \Rightarrow \lambda^2 = \frac{13}{16} \Rightarrow \lambda = \pm\frac{\sqrt{13}}{4}$. Thus, $\left(\pm\frac{4}{\sqrt{13}}, \pm\frac{6}{\sqrt{13}}\right)$ are the required extreme points and value of the function f is $\pm 2\sqrt{13}$.

Note: Similarly, one can solve problems having two constraints here $\nabla f = \lambda \nabla g_1 + \mu \Delta g_2$, $g_1(x, y, z) = 0$ and $g_2(x, y, z) = 0$ are to be solved simultaneously for independent variables x, y and z and Lagrange multipliers λ and μ.

6.9 TAYLOR'S FORMULA FOR TWO VARIABLES

In this section, we discuss application of Taylor's formula to derive the second derivative test for local extreme values and the error formula for linearization of function of two variables.

Second Derivative Test

Let $f(x, y)$ have continuous partial derivatives in the neighborhood of a point $P(a, b)$ where $f_x = f_y = 0$. Take a point $Q(a + h, b + k)$ with h and k as very small increment. We parameterize line PQ as $x = a + th, y = b + tk, 0 \le t \le 1$.

If $F(t) = f(a + th, b + tk)$ then by chain rule $F'(t) = f_x \frac{dx}{dt} + f_y \frac{dy}{dt} = nf_x + kf_y$.

Given f_x and f_y are differentiable. This makes F' differentiable in t and

$$F''(t) = \frac{\partial F'}{\partial x}\frac{dx}{dt} + \frac{\partial F'}{\partial x}\frac{\partial y}{\partial t} = \frac{\partial}{\partial x}(hf_x + kf_y)h + \frac{\partial}{\partial y}(hf_x + kf_y)k$$

$$= h^2 f_{xx} + 2hk\, f_{xy} + k^2 f_{yy}$$

Since F and F' are continuous on [0, 1] and F' is differentiable on (0, 1), we can apply Taylor's Formula with $n = 2$ and $a = 0$ to have

$$F(1) = F(0) + F'(0)(1-0) + F''(c)\frac{(1-0)^2}{2}$$

$$\Rightarrow F(1) = F(0) + F'(0) + \frac{1}{2}F''(c),\ c \in (0,1).$$

So,

$$f(a+h, b+k) = f(a, b) + hf_x(a, b) + kf_y(a, b) + \frac{1}{2}(h^2 f_{xx} + 2hk\, f_{xy} + k^2 f_{yy})\,|_{(a+ch, b+ck)}$$

Since, $f_x(a, b) = f_y(a, b) = 0$, this reduces to

$$f(a+h, b+k) = f(a, b) + \frac{1}{2}(h^2 f_{xx} + 2hk\, f_{xy} + k^2 f_{yy})\,|_{(a+ch,\, b+ck)}$$

An extreme value of f at (a, b) can be determined by the sign of $f(a + h, b + k) - (a, b)$ which will be same as the sign of $X(c) = (h^2 f_{xx} + 2hk\, f_{xy} + k^2 f_{yy})\,|_{(a+ch, b+ck)}$. Now, if $Q(0) \neq 0$, the sign of $Q(c)$ will be same as the sign of $Q(0) = h^2 f_{xx}(a, b) + 2hk\, f_{yy}\,(a, b) + k^2 f_{yy}(a, b)$ for small values of h and k. The sign of $Q(0)$ can be observed from the sign of f_{xx} and $f_{xx} f_{yy} - f_{xy}^2$ at (a, b). Multiply both sides of $Q(0)$ by f_{xx}. Then we get, $f_{xx}Q(0) = (hf_{xx} + kf_{xy})^2 + (f_{xx}f_{yy} - f_{xy}^2)k^2$.

Observations:

1. If $f_{xx} < 0$ and $f_{xx} f_{yy} - f_{xy}^2 > 0$ at (a, b) then $Q(0) < 0$ for all small values of h and k, and f has a local maximum value at (a, b).
2. If $f_{xx} > 0$ and $f_{xx} f_{yy} - f_{xy}^2 > 0$ and at (a, b) then $Q(0) > 0$ for all small values of h and k and f has a local minimum values at (a, b).
3. If $f_{xx} f_{yy} - f_{xy}^2 < 0$ at (a, b), then there can be $Q(0) > 0$ and $Q(0) < 0$ for small values of h and k. We end up with conclusion that f has a saddle point at (a, b).
4. Test fails when $f_{xx} f_{yy} - f_{xy}^2 = 0$.

The Error Formula for Linear Approximations

We want to prove that the difference $E(x, y)$, between the values of a function $f(x, y)$ and its linearization $L(x, y)$ at (x_0, y_0) satisfies the inequality $|E(x,y)| \leq \frac{1}{2}M(|x-x_0| + |y-y_0|)^2$. Here, M is the upper bound for $|f_{xx}|$, $|f_{yy}|$ and $|f_{xy}|$ on R, a closed rectangular region centered at (x_0, y_0). We substitute x_0 for a, y_0 for b and x–x_0 and y–y_0 for h and k, respectively. We get

$$f(x,y) = \underbrace{f(x_0,y_0) + f_x(x_0,y_0)(x-x_0) + f_y(x_0,y_0)(y-y_0)}_{\text{linearization } L(x,y)}$$

$$+ \underbrace{\frac{1}{2}((x-x_0)^2 f_{xx} + 2(x-x_0)(y-y_0)f_{xy} + (y-y_0)^2 f_{yy}(x_0 + c(x-x_0), y_0 + c(y-y_0)))}_{\text{Error } E(x,y)}$$

$$\Rightarrow |E| \le \frac{1}{2}\left(|x-x_0|^2 + 2|x-x_0||y-y_0||f_{xy}| + |y-y_0|^2|f_{yy}|\right)$$

Now, $|f_{xx}| \le M$, $|f_{xy}| \le M$, and $|f_{yy}| \le M$ on R.

So, $|E| = \frac{1}{2}M\left(|x-x_0| + |y-y_0|\right)^2$.

Taylor Formulas for Function of Two Variables

We have seen that F' and F'' can be obtained by differentiating partially $f(x, y)$ as:

$$\left(h\frac{\partial}{\partial x} + k\frac{\partial}{\partial y}\right) \text{ and } \left(h\frac{\partial}{\partial x} + k\frac{\partial}{\partial y}\right)^2 = h^2\frac{\partial^2}{\partial x^2} + 2hk\frac{\partial^2}{\partial xy} + k^2\frac{\partial^2}{\partial y^2}$$

In general, $F^{(n)}(t) = \frac{d^n}{dt^2}F(t) = \left(h\frac{\partial}{\partial x} + k\frac{\partial}{\partial y}\right)^n + (x, y)$

This suggests and applying $\frac{d^n}{dt^n}$ to $F(t)$ gives the operator $\left(h\frac{\partial}{\partial x} + k\frac{\partial}{\partial y}\right)^n$ to $f(x, y)$ after expanding it by the Binomial Theorem.

It $(n + 1)$ order partial derivatives of f are continuous in a neighborhood of (a, b) then $F(t) = F(0) + F'(0)t + F''(0)\frac{t^2}{2!} + \ldots + \frac{F^{(n)}(0)}{n!}t^n + \text{remainder}$

At $t = 1$, we obtain

$$F(1) = F(0) + F'(0) + \frac{F''(0)}{2!} + \ldots + \frac{F^{(n)}(0)}{n!} + \text{remainder}$$

This will help us to derive Taylor's formula for $f(x, y)$ at (a, b).

Let $(n + 1)$–order partial derivatives of $f(x, y)$ are continuous in the neighborhood of (a, b), then

$$f(a+h, b+k) = f(a, b) + \left(h\frac{\partial}{\partial x} + k\frac{\partial}{\partial y}\right)f\Big|_{(a,b)} + \frac{1}{2!}\left(h\frac{\partial}{\partial x} + k\frac{\partial}{\partial y}\right)^2 f\Big|_{(a,b)} + \ldots$$
$$+ \frac{1}{n!}\left(h\frac{\partial}{\partial x} + k\frac{\partial}{\partial y}\right)^n f\Big|_{(a,b)} + \frac{1}{(n+1)!}\left(h\frac{\partial}{\partial x} + k\frac{\partial}{\partial y}\right)^{n+1} f\Big|_{(a+ch,b+ck)}$$

EXAMPLE 6.41 Find a quadratic approximation to $f(x,y) = e^{x+y}$ near to the origin. What is the approximate value it $|x| \le 0$ and $|y| \le 0.1$?

Ans: We want quadratic approximation of given function at (0,0). So by Taylor's formula,

$$f(x,y) = f(0,0) + (xf_x + yf_y)(0,0) + \frac{1}{2}(x^2 f_{xx} + 2xyf_{xy} + y^2 f_{yy})(0,0)$$
$$+ \frac{1}{6}(x^3 f_{xxx} + 3x^2 f_{xxy} + 3xy^2 f_{yy} + y^3 f_{yy})(0+h, 0+k)$$

We have $f(0, 0) = 1, f_x(0, 0) = e^{x+y}\Big|_{(0,0)} = 1$ and $f_y(0, 0) = e^{x+y}\Big|_{(0,0)} = 1$

$$f_{xx}(0,0) = f_{xy}(0,0) = f_{yy}(0,0) = 1$$

Then, $f(x,y) \approx 1 + (x+y) + \frac{1}{2}(x+y)^2$.

The error in approximation is $E(x,y) = \frac{1}{6}(x+y)^3\Big|_{(cx,\, cy)}$

Hence, $|E(x,y)| \le \frac{1}{6}(0.2)^3 = \frac{8}{6}(0.001) \approx 0.00134$. The error will not exceed 0.00134 if $|x| \le 0.1$ and $|y| \le 0.1$.

EXERCISES

Exercises based on Section 6.1:

1. Find value of the function $f(x, y) = x^2 + xy^3$ at $f(0, 0)$, $f(-1, -1)$, $f(2, 3)$.
2. Find value of the function $f(x,y,z) = \dfrac{x-y}{y^2+z^2}$ at $f(2, 2, 10), f\left(1, \frac{1}{2}, \frac{1}{4}\right)$.
3. Find $f(0, 0, 0)$ and $f(2, -3, 6)$ where $f(x,y,z) = \sqrt{49 - x^2 + y^2 + z^2}$.

Exercises based on Section 6.2:

4. Find $\displaystyle\lim_{(x,y)\to(0,0)} \frac{x^2 - 3y^2 + 5}{x^2 + y^2 + 4}$.
5. Find $\displaystyle\lim_{(x,y)\to\left(0, \frac{\pi}{4}\right)} \sec x \tan y$.
6. Find $\displaystyle\lim_{(x,y)\to\left(1, \frac{\pi}{6}\right)} \frac{x \sin y}{x^2 + 1}$.
7. Find $\displaystyle\lim_{(x,y)\to(3,4)} \sqrt{x^2 + y^2 - 1}$.

8. Find $\lim\limits_{(x,y)\to(0,0)} \dfrac{e^x \sin x}{x}$.

9. Find $\lim\limits_{(x,y)\to(0,0)} \dfrac{x-y+2\sqrt{x}-2\sqrt{y}}{\sqrt{x}-\sqrt{y}}, x \neq y$.

10. Find $\lim\limits_{(x,y)\to(1,1)} \dfrac{xy-y+2x+2}{x-1}, x \neq 1$.

11. Find $\lim\limits_{(x,y)\to(2,0)} \dfrac{\sqrt{2x-y-2}}{2x-y-4}, 2x-y \neq 4$.

12. Find $\lim\limits_{(x,y)\to(1,-1)} \dfrac{x^3+y^3}{x+y}$.

13. Find $\lim\limits_{(x,y)\to(0,0)} \dfrac{\sin(x^2+y^2)}{x^2+y^2}$.

14. Find $\lim\limits_{(x,y,z)\to(\pi,\pi,0)} (\sin^2 x + \cos^2 y + \sec^2 z)$.

15. Find $\lim\limits_{(x,y,z)\to(1,-1,-1)} \dfrac{2xy+yz}{x^2+z^2}$.

16. Find $\lim\limits_{(x,y,z)\to(\pi,0,3)} ze^{-2y} \cos 2x$.

17. Find $\lim\limits_{(x,y,z)\to(2,3,-6)} \ln\sqrt{x^2+y^2+z^2}$.

18. At what points (x, y) in the plane for which $f(x,y) = \sin\dfrac{1}{xy}$ is continuous?

19. At what points (x, y) in the plane for which the function $f(x, y, z) = \ln(z - x^2 - y^2 - 1)$ is continuous?

20. Show that the function $f(x,y) = \dfrac{-x}{\sqrt{x^2+y^2}}$ have no limit as $(x, y) \to (0, 0)$

21. Show that the function $\lim\limits_{(x,y)\to(0,0)} \dfrac{x^4-y^2}{x^4+y^2}$ has no limit.

22. Show that the function $\lim\limits_{(x,y)\to(0,0)} \dfrac{x-y}{x+y}$ has no limit.

23. Let $f(x,y) = \begin{cases} 1, y \geq x^2 \\ 1, y \leq 0 \\ 0, \text{otherwise} \end{cases}$, find

(a) $\lim_{(x,y)\to(0,1)} f(x,y)$

(b) $\lim_{(x,y)\to(2,3)} f(x,y)$

(c) $\lim_{(x,y)\to(0,0)} f(x,y)$

24. Let $f(x,y) = \begin{cases} x^2, x \geq 0 \\ x^3, x \leq 0 \end{cases}$. Find

(a) $\lim_{(x,y)\to(3,-2)} f(x,y)$

(b) $\lim_{(x,y)\to(-2,1)} f(x,y)$

(c) $\lim_{(x,y)\to(0,0)} f(x,y)$

25. Knowing that $1 - \frac{x^2y^2}{3} < \frac{\tan^{-1} xy}{xy} < 1$, find $\lim_{(x,y)\to(0,0)} \frac{\tan^{-1} xy}{xy}$.

26. Knowing that

$$2|xy| - \frac{x^2y^2}{6} < 4 - 4\cos\sqrt{|xy|} < 2|xy|, \text{ find } \lim_{(x,y)\to(0,0)} \frac{4 - 4\cos\sqrt{|xy|}}{|xy|}$$

27. Find $\lim_{(x,y)\to(0,0)} \tan^{-1}\left(\frac{|x| + |y|}{x^2 + y^2}\right)$.

28. Find $\lim_{(x,y)\to(0,0)} \frac{x^3 - xy^2}{x^2 + y^2}$.

29. Find $\lim_{(x,y)\to(0,0)} \frac{y^2}{x^2 + y^2}$.

30. Define $f(0, 0)$ to make $f(x,y) = \ln\left(\frac{3x^2 - x^2y^2 + 3y^2}{x^2 + y^2}\right)$ continuous.

31. Define $f(0, 0)$ so that $f(x,y) = \frac{2x^2y}{x^2 + y^2}$ is continuous.

32. Determine δ for which $f(x,y) = x^2 + y^2$, $\varepsilon = 0.01$ is continuous.

33. Determine δ for which $f(x, y) = \dfrac{xy^2}{x^2 + y^2}$ and $f(0, 0) = 0$, $\varepsilon = 0.04$ is continuous.

34. Given $f(x, y) = \dfrac{x + y}{x^2 + 1}$ and $\varepsilon = 0.01$ to be continuous find δ.

35. Determine δ for which $f(x, y, z) = x^2 + y^2 + z^2$, and $\varepsilon = 0.015$.

36. Determine δ for which $f(x, y, z) = \dfrac{x + y + z}{x^2 + y^2 + z^2 + 1}$ and $\varepsilon = 0.015$.

Exercises based on Section 6.3:

Find first order partial derivatives, f_x and f_y of f

37. $f(x, y) = x^2 - xy + y^2$.
38. $f(x, y) = (x^2 - 1)(y + 2)$.
39. $f(x,y) = 3xy - 4x^2 - y^2 + 5x - 6y + 8$.
40. $f(x, y) = (xy - 4)^2$.
41. $f(x,y) = \sqrt{x^3 + y^3}$.
42. $f(x,y) = \tan^{-1}\left(\dfrac{y}{x}\right)$.
43. $f(x,y) = \dfrac{1}{(x + y)}$.
44. $f(x,y) = e^{-x}\sin(x + y)$.
45. $f(x,y) = e^{x+y+1}$.
46. $f(x,y) = \ln(x + y)$.
47. $f(x,y) = x^y$
48. $f(x,y) = \int_x^y g(t)\,dt$ (g is continuous for all t).

Find f_x, f_y, f_z

49. $f(x,y,z) = xy + yz + zx$
50. $f(x, y, z) = (x^2 + y^2 + z^2)^{-\frac{1}{2}}$
51. $f(x,y,z) = \sin^{-1}(xyz)$
52. $f(x, y, z) = \ln(x + 2y + 3z)$

53. $f(x, y, z) = \tan h\,(x + 2y + 3z)$

54. $f(x, \alpha) = \cos(2\pi x - \alpha)$

55. $f(r, \phi, \theta) = r \sin \phi \cos \theta$

56. Work done by the heart is given by $w(P, V, \delta, v, g) = PV + \dfrac{V\delta v^2}{2g}$.

    Find rate of change of work done with respect to each variable.

Find all the Second-order Partial Derivatives:

57. $f(x,y) = x + y + xy$

58. $f(x,y) = x^2 y + \cos y + y \sin x$

59. $f(x,y) = \ln(x + y)$

60. $f(x,y) = x \sin(x^2 y)$

61. For $f(x,y) = e^x + x \ln y + y \ln x$, verify that $f_{xy} = f_{yx}$

62. For $f(x,y) = xy^2 + x^2 y^3 + x^3 y$, verify that $f_{xy} = f_{yx}$

63. Evaluate $f_x(1, 2)$ and $f_y(1, 2)$ for $f(x,y) = 1 - x + y - 3x^2 y$.

64. Evaluate $f_x(-2, 3)$ and $f_y(-2, 3)$ if $f(x,y) = \sqrt{2x + 3y - 1}$.

65. Define $f(x,y) = 2x + 3y - 5$. Find the slope of the line tangent to this surface at the point (2, –1) and lying in the (a) plane $x = 2$ and (b) plane $y = -1$.

66. Find the value of $\dfrac{\partial z}{\partial x}$ at the point (1,1,1) if the equation $xy + z^3 x - 2yz = 0$ defines z as a function of two independent variables x and y and the partial derivatives exists.

Exercises based on Section 6.4:

Find $\dfrac{dw}{dt}$ and evaluate it at a given point.

67. $w = x^2 + y^2,\ x = \cos t,\ y = \sin t,\ t = \pi.$

68. $w = 2ye^x - \ln z,\ x = \ln(t^2 + 1),\ y = \tan^{-1} t,\ z = e^t,\ t = 1.$

69. Find z_u and z_v if $z = 4e^x \ln y,\ x = \ln(u \cos v),\ y = u \sin v,\ (u, v) = \left(2, \dfrac{\pi}{4}\right)$

70. Find w_u and w_v and its values at $(u, v) = \left(\dfrac{1}{2}, 1\right)$ if $w = xy + yz + zx,\ x = u + v,\ y = u - v,\ z = uv$

71. Find u_x, u_y and u_z and its values at the specified point. If $u=\dfrac{p-q}{q-r}, p=x+y+z, q=x-y+z, r=x+-z,(x,y,z)=(\sqrt{3},2,1)$.

72. Find $\dfrac{dy}{dx}$ at $(1,1)$ if $x^3-2y^2+xy=0$.

73. For $x^2+xy+y^2-15=0$, find $\dfrac{dy}{dx}$ at $(1,2)$.

74. For $z^3-xy+yz+y^3-1=0$, find $\dfrac{\partial z}{\partial x}$ and $\dfrac{\partial z}{\partial y}$ at $(1,1,1)$.

75. Given $\sin(x+y)+\sin(y+z)+\sin(x+z)=0$. Compute $\dfrac{\partial z}{\partial x}$ and $\dfrac{\partial z}{\partial y}$ at (π, π, π).

76. Find $\partial w/\partial r$ when $r = 1$, $s = -1$ if $w = (x + y + z)^2$, $x = r - s$, $y = \cos(r + s)$, $z = \sin(r + s)$

77. Find $\partial w/\partial v$ when $u = 0$, $v = 0$ if $w=x^2+\dfrac{y}{x}$, $x = u - 2v + 1$, $y = 2u + v - 2$.

78. Find $\dfrac{\partial z}{\partial u}$ and $\dfrac{\partial z}{\partial v}$ when $u = \ln 2$, $v = 1$ if $z = 5\tan^{-1}x$ and $x = e^u + \ln v$.

79. **Extreme values on a helix.** Suppose that the partial derivatives of a function $f(x, y, z)$ at points on the helix $x = \cos t$, $y = \sin t$, $z = t$ are $f_x = \cos t$, $f_y = \sin t$, $f_z = t^2 + t - 2$. At what points on the curve, if any, can f take on execute values?

80. **Temperature on a circle.** Let $T = f(x, y)$ be the temperature on the circle $x = \cos t$, $y = \sin t$, $0 \le t \le 2\pi$ and support that $\dfrac{\partial T}{\partial x}=8x-4y$, $\dfrac{\partial T}{\partial x}=8y-4x$. Find where the maximum and minimum temperature on the circle occur by examining the derivatives dT/dt, and $\dfrac{d^2T}{dt^2}$.

Exercises based on Section 6.5:

81. Find ∇f at (1,1,1) if $f(x,y,z)=x^2+y^2-2z^2+z\ln x$.

82. Find ∇f at (–1, 2, –2) if $f(x,y,z)=(x^2+y^2+z^2)^{-\frac{1}{2}}+\ln(xyz)$.

83. Find the derivative of $f(x,y)=2xy-3y^2$ at $P(5, 5)$ in the direction of $\vec{n}=4\hat{i}+3\hat{j}$.

84. Find the derivative of $f(x,y)=\dfrac{x-y}{xy+2}$ at $P(1, -1)$ in the direction of $\vec{n}=12\hat{i}+5\hat{j}$.

85. Find the derivative of $f(x, y, z) = 3e^x \cos yz$ at $P(0, 0, 0)$ in the direction of $\vec{n} = 2\hat{i} + \hat{j} - 2\hat{k}$.
86. Find the direction in which the function $f(x, y) = x^2 + xy + y^2$ increases and decreases most rapidly at $P(-1, 1)$. Then find the derivatives of the function in these directions.
87. Find the direction in which the function $f(x, y, z) = \ln xy + \ln yz + \ln zx$ increases and decreases most rapidly at (1,1,1). Find the derivatives of the function in these directions.
88. Let $f(x, y) = x^2 + y^2 - xy - y$. Find the directions $\vec{n}$ and the values of $D_{\hat{n}} f(1, -1)$ for which
    (a) $D_{\hat{n}} f(1, -1)$ is largest.
    (b) $D_{\hat{n}} (1, -1)$ is smallest.
    (c) $D_{\hat{n}} f(1, -1) = 0$
    (d) $D_{\hat{n}} f(1, -1) = 4$

Exercises based on Section 6.6:

89. Find the equation for tangent plane and normal line at the point $P(1,1,1)$ on the surface $x^2 + y^2 + z^2 = 3$.
90. Find the equation for tangent plane and normal line at the point $P(0,1,0)$ on the surface $x + y + z = 1$.
91. Find the equation of the tangent plane at (1,0,0) to the surface $z = \ln(x^2 + y^2)$.
92. Find the equation of the tangent plane at (1,2,1) to the surface $z^2 = y - x$.
93. Find parametric equation for the line tangent to the curve of intersection of surfaces $x + y^2 + 2z = 4, x = 1$ at point (1,1,1).
94. Find parametric equation for the line tangent to the curve of intersection of surfaces $x^3 + 3x^2y^2 + y^3 + 4xy - z^2 = 0, x^2 + y^2 + z^2 = 11$ at a point (1,1,3).
95. What will be the change in $f(x, y, z) = \ln\sqrt{x^2 + y^2 + z^2}$ if the point $P(x, y, z)$ moves from $Q(3, 4, 12)$ a distance of $dh = 0.1$unit in the direction of $3\hat{i} + 6\hat{j} - 2\hat{k}$?
96. **Temperature change along a circle.** Suppose that Celsius temperature at a point (x, y) in the xy-plane is $T(x,y) = x \sin 2y$ and that distance in the xy-plane is measured in meters. A particle is moving clockwise around the unit circle at the constant rate of 2 m/sec. How fast is the temperature experienced by the particle changing in the degree Celsius per meter at the point $P\left(\frac{1}{2}, \frac{\sqrt{3}}{2}\right)$?

97. Find the linearization $L(x, y)$ of the function $f(x,y)=x^2+y^2+1$ at (0,0) and (1,1).
98. Find the linearization $L(x, y)$ of the function $f(x,y)=3x+4y-5$ at (0,0) and (1,1).
99. Find the linearization $L(x, y)$ of the function $f(x,y)=e^x \cos y$ at $\left(0,\frac{\pi}{2}\right)$.
100. Find the linearization $L(x, y)$ of the function $f(x,y)=x^2-3xy+5$ at (2,1) and also find an upper bound for the magnitude $|E|$ over the rectangle $R:|x-2|\le 0.1,\ |y-1|\le 0.1$.
101. Find the linearization of the function $f(x,y)=1+y+x\cos y$ at (0, 0) and also find an upper bound for the magnitude $|E|$ over the rectangle $R:|x|\le 0.2,\ |y|\le 0.2$.
102. Find the linearization $L(x,y,z)$ of the function $f(x,y,z)=xy+yz+xz$ at (1,1,1), (1,0,0) and (0,0,0).
103. Find the linearization $L(x,y,z)$ of the function $f(x,y,z)=\sqrt{x^2+y^2+z^2}$ at (1,0,0), (1,1,0) and (1,2,2).
104. Find the linearization $L(x,y,z)$ of the function $f(x,y,z)=xy+2yz-3xz$ at (1,1,0) and also found upper bound for the magnitude $|E|$ over the rectangle $R:|x-1|\le 0.01,\ |y-1|\le 0.01,\ |z|\le 0.01$.

Exercises based on Section 6.7:

105. Find extreme values of $f(x,y)=x^2+y^2+xy+3x-3y+4$.
106. Find extreme values of $f(x,y)=2x^2+3xy+4y^2-5x+2y$.
107. Find extreme values of $f(x,y)=x^2+xy+3x+2y+5$.
108. Find extreme values of $f(x,y)=x^2-y^2-2x+4y+6$.
109. Find extreme values of $f(x,y)=2xy-x^2-2y^2+3x+4$.
110. Find extreme values of $f(x,y)=2\ln x+\ln y-4x-y$.
111. Find extreme values of $f(x,y)=e^{x^2+y^2-4x}$.
112. Find extreme values of $f(x, y) = y \sin x$.
113. Find two numbers a and b with $a \le b$ such that $\int_a^b (b-x-x^2)\,dx$ has its largest value.
114. Among all the points on the graph $z=10-x^2-y^2$ that lie above the plane $x+2y+3z=0$, find the point farthest form the plane.
115. Find the point on the plane $3x+2y+z=6$ that is nearest the origin.
116. Find three numbers whose sum is 9 and whose sum of square is minimum.

117. Find the maximum value of $r = xy + yz + xz$, where $x + y + z = 6$.
118. Find the dimension of the rectangle box of maximum value that can be inscribed inside the sphere $x^2 + y^2 + z^2 = 4$.

Exercises based on 6.8:

119. Find the maximum value of $f(x,y) = 49 - x^2 - y^2$ on the line $x + 3y = 10$.
120. Find the points on the curve $xy^2 = 54$ nearest the origin.
121. Find the minimum values of $x + y$, subject to the constraint $xy = 16$, $x > 0$, $y > 0$.
122. Find the maximum value of xy subject to constraint $x + y = 16$.
123. Find the dimensions of the rectangle of greatest area that can be inscribed in the ellipse $\frac{x^2}{16} + \frac{y^2}{9} = 1$ with sides parallel to coordinate axes.
124. Find the dimension of the closed right circular cylindrical can of smallest surface area whose volume is $16\pi \text{cm}^3$.
125. Find the minimum distance from the surface $x^2 - y^2 - z^2 = 1$ to the (0,0,0).
126. Find the maximum and minimum values of $f(x,y,z) = x - 2y + 5z$ on the sphere $x^2 + y^2 + z^2 = 30$.
127. A space probe in the shape of the ellipsoid $4x^2 + y^2 + 4z^2 = 16$ enters earth atmosphere and its surface begins to heat. After one hour, the temperature at the point (x, y, z) on the probe's surface is $T = (x,y,z) = 8x^2 + 4yz - 16z + 600$. Find the hottest point on the probe's surface.

Exercises based on Section 6.9:

128. Find quadratic and cubic approximation of $f(x, y) = xe^y$ near the origin.
129. Find quadratic and cubic approximation of $f(x,y) = y \sin x$ near the origin.
130. Find quadratic and cubic approximation of $f(x,y) = e^x \ln(1+y)$ near the origin.

Theoretical Questions

131. What is real-valued function of two independent variables? Three independent variables? Give examples.
132. Define a limit of a function of two variables. What are the basic properties of limits of function of two independent variables?
133. When a function of two variables is said to be continuous?

134. What can be said about algebraic combinations and composites of continuous functions?
135. State the two path test for non-existence of limits.
136. Define the partial derivation $\frac{\partial f}{\partial x}$ and $\frac{\partial f}{\partial y}$ of a function $f(x, y)$.
137. What is the Mixed Derivative Theorem for mixed second order partial derivations?
138. State chain rule for function of two independent variables, three independent variables, and function defined on surface. Explain using diagrams.
139. What is the gradient vector of a differentiable function $f(x, y)$? How it is related to the functions directional derivatives?
140. How can you use directional derivation to estimate change?
141. What can you say about the accuracy of linear approximation of function of two independent variables?
142. Define local maxima, local minima and saddle points for a differentiable function $f(x, y)$.
143. State second derivatives test to characterize extreme values.
144. Explain the method of Lagrange multiplies and gives examples.
145. How does Taylor's formula for a function $f(x, y)$ generate polynomial approximations and error estimates?

Answers:

1. 0, 0, 58
2. $0, \frac{8}{5}$
3. 7, 0
4. $\frac{5}{4}$
5. 1
6. $\frac{1}{4}$
7. $2\sqrt{6}$
8. 1
9. 2
10. –1
11. $\frac{1}{4}$
12. 3
13. 1
14. 2
15. $-\frac{1}{2}$
16. 3
17. ln 7
18. All except $x = 0, y = 0$
19. Except $z > x^2 + y^2 + 1$
20. *Hint:* Consider the path $y = x, x > 0$ and $y = x, x < 0$
21. *Hint:* Consider the path $y = mx^2$
22. *Hint:* Consider the path $y = mx$
23. 1, 0, does not exist
24. 9, –8, 0
25. 1

26. 2

27. $\pi/2$

28. 0

29. Limit does not exist

30. $f(0,0) = \ln 3$

31. $f(0,0) = 0$

32. $\delta = 0.1$

33. $\delta = 0.04$

34. $\delta = 0.005$

35. $\delta = \sqrt{0.015}$

36. $\delta = 0.0005$

37. $f_x = 2x - y,$ $f_y = 2y - x$

38. $f_x = 2x(y + 2), f_y = x^2 - 1$

39. $f_x = 3y - 8x + 5, f_y = 3x - 2y - 6$

40. $f_x = 2y(xy - 4), f_y = 2x(xy - 4)$

41. $f_x = \dfrac{3x^2}{2\sqrt{x^3 + y^3}}, f_y = \dfrac{3y^2}{2\sqrt{x^3 + y^3}}$

42. $f_x = \dfrac{-y}{x^2 + y^2}, f_y = \dfrac{x}{x^2 + y^2}$

43. $f_x = -\dfrac{1}{(x + y)^2}, f_y = -\dfrac{1}{(x + y)^2}$

44. $f_x = e^{-x}(\cos(x + y) - \sin(x + y)), f_y = e^{-x}\cos(x + y)$

45. $f_x = e^{x+y+1}, f_y = e^{x+y+1}$

46. $f_x = \dfrac{1}{x + y}, f_y = \dfrac{1}{x + y}$

47. $f_x = yx^{y-1}, f_y = x^y \ln x$

48. $f_x = -g(x), f_y = g(y)$

49. $f_x = y + z, f_y = x + z, f_z = x + y$

50. $f_x = -x(x^2 + y^2 + z^2)^{-\frac{3}{2}}, f_y = -y(x^2 + y^2 + z^2)^{-\frac{3}{2}}, f_z = -z(x^2 + y^2 + z^2)^{-\frac{3}{2}}$

51. $f_x = \dfrac{yz}{\sqrt{1 - (xyz)^2}}, f_y = \dfrac{xz}{\sqrt{1 - (xyz)^2}}, f_z = \dfrac{xy}{\sqrt{1 - (xyz)^2}}$

52. $f_x = \dfrac{1}{x + 2y + 3z}, f'_y = \dfrac{2}{x + 2y + 3z}, f_z = \dfrac{3}{x + 2y + 3z}$

53. $f_x = \sec^2(x + 2y + 3z), f_y = 2\sec^2(x + 2y + 3z), f_z = 3\sec^2(x + 2y + 3z)$

54. $f_\alpha = -2\pi \sin(2\pi x - \alpha), f_x = \sin(2\pi x - \alpha)$

55. $f_r = \sin\phi \cos\theta, f_\phi = r\cos\phi \cos\theta, f_\theta = -r\sin\phi \sin\theta$

56. $w_p = V, w_V = P + \dfrac{\delta v^2}{2g}, w_\delta = \dfrac{Vv^2}{2g}, w_v = \dfrac{V\delta v}{g}, w_g = -\dfrac{V\delta v^2}{2g^2}$

57. $f_{xx} = 0, f_{xy} = f_{yx} = 1, f_{yy} = 0$

58. $f_{xx} = 2y - y\sin x, f_{yy} = -y\cos x, f_{xy} = f_{yx} = 2x + \cos x$

59. $f_{xx} = \dfrac{-1}{(x+y)^2}, f_{yy} = \dfrac{-1}{(x+y)^2}$

60. $f_{xx} = 6\,xy\cos(x^2y), f_{yy} = -x^5\sin(x^2y), f_{xy} = f_{yx} = 3x^2\cos(x^2y) - 2x^4y\sin(x^2y)$

63. –13, –2 64. $\dfrac{1}{2}, \dfrac{3}{4}$ 65. 3, 2 66. –2

67. 0, 0 68. $1 + 4t\tan^{-1}t,\ \pi + 1$

69. $z_u = 4\cos v\ \ln(u\sin v) + 4\cos v,\ \sqrt{2}(\ln 2 + 2)$

$z_v = -4u\sin v\ \ln(u\sin v) + \dfrac{4u\cos^2 v}{\sin v},\ -2\sqrt{2}(\ln 2 - 2)$

70. $w_u = 3$ and $w_v = 3/2$

71. $u_x = 0, u_y = \dfrac{z}{(z-y)^2}, u_z = \dfrac{-y}{(z-y)^2}$, 0, 1, –2

72. $\dfrac{4}{3}$ 73. $-\dfrac{4}{5}$ 74. $\dfrac{1}{4}, -\dfrac{3}{4}$ 75. –1, –1

76. 12 77. –7 78. $\dfrac{\partial z}{\partial u} = 2, \dfrac{\partial z}{\partial v} = 1$

79. $(\cos(1), \sin(1), 1), (\cos(-2), \sin(-2), -2)$

80. Maximum at $\left(-\dfrac{\sqrt{2}}{2}, \dfrac{\sqrt{2}}{2}\right)$ and $\left(\dfrac{\sqrt{2}}{2}, -\dfrac{\sqrt{2}}{2}\right)$,

Minimum at $\left(\dfrac{\sqrt{2}}{2}, \dfrac{\sqrt{2}}{2}\right)$ and $\left(-\dfrac{\sqrt{2}}{2}, -\dfrac{\sqrt{2}}{2}\right)$

81. $3\hat{i} + 2\hat{j} - 4\hat{k}$ 82. $\dfrac{1}{54}(-52\hat{i} + 23\hat{j} - 23\hat{k})$

83. –4 84. $\dfrac{21}{13}$ 85. 2

86. Increases rapidly in the direction $\vec{n} = \dfrac{1}{3\sqrt{3}}(\hat{i} - 5\hat{j} - \hat{k}), (D_{\vec{n}} f)_p = 3\sqrt{3}$

Decreases rapidly in the direction $-\vec{n} = \dfrac{1}{3\sqrt{3}}(-\hat{i} + 5\hat{j} + \hat{k}), (D_{\vec{n}} f)_p = -3\sqrt{3}$

87. Increases rapidly in the direction $\vec{n} = \dfrac{1}{\sqrt{3}}(\hat{i} + \hat{j} + \hat{k}), (D_{\vec{n}} f)_p = 2\sqrt{3}$

88. (a) 5, (b) 5, (c) $\hat{n} = \frac{1}{5}(4\hat{i} + 3\hat{j}),\ \frac{-1}{5}(4\hat{i} + 3\hat{j})$,

(d) $\hat{n} = -\hat{j},\ \hat{n} = \frac{1}{25}(24\hat{i} - 7\hat{j})$

89. Tangent plane: $x + y + z = 3$, Normal line: $x = 1 + 2t$, $y = 1 + 2t$, $z = 1 + 2t$.

90. Tangent plane: $x + y + z = 1$ normal line: $x = t$, $y = 1 + t$, $z = t$

91. $2x - z - 2 = 0$

92. $x - y + 2z - 1 = 0$

93. $x = 1, y = 1 + 2t, z = 1 - 2t$

94. $x = 1 + 90t$, $y = 1 - 90t$, $z = 3$

95. $df = \frac{9}{11830}$

96. $\frac{\sqrt{3}}{2}\sin\sqrt{3} - \frac{1}{2}\cos\sqrt{3} \approx 0.935°\text{c/m}$

97. $1, 2x + 2y - 1$

98. $3x + 4y - 5, 3x + 4y - 5$

99. $1 + x - y + \frac{\pi}{2}$

100. $1 + x$, 0.0222

101. $x + y + 1$, 0.08

102. $2x + 2y + 2z - 3, y + z, 0$

103. $x,\ \frac{1}{\sqrt{2}}x + \frac{1}{\sqrt{2}}y,\ \frac{1}{3}x + \frac{2}{3}y + \frac{2}{3}z$

104. $x + y - z = 1, 0.00135$

105. Local minimum $(-3, 3)$, $f(-3, 3) = -5$

106. Local minimum $(2, -1)$, $f(2, -1) = -6$

107. Saddle point $(-2, 1)$

108. Saddle point $(1, 2)$

109. $\left(3, \frac{3}{2}\right), f\left(3, \frac{3}{2}\right) = \frac{17}{2}$

110. Local maximum $\left(\frac{1}{2}, 1\right), f\left(\frac{1}{2}, 1\right) = \ln\left(\frac{1}{4}\right) - 3$

111. Local minimum $(2,0)$, $f(2,0) = e^{-4}$

112. Saddle point $(\pi, \pi, 0)$

113. $a = -3$, $b = 2$

114. $\left(\frac{1}{6}, \frac{1}{3}\right), \frac{355}{36}$

115. $\left(\frac{9}{7}, \frac{6}{7}, \frac{3}{7}\right)$

116. (3,3,3)

117. 12

118. $\frac{4\sqrt{3}}{3} \times \frac{4\sqrt{3}}{3} \times \frac{4\sqrt{3}}{3}$

119. 39

120. $\left(3, \frac{3}{\sqrt{2}}\right)$

121. 8 122. 64 123. $4\sqrt{3} \times 3\sqrt{2}$

124. $r = 2$, $h = 4$ 125. 1

126. Maximum value of $f(1, -2, 5) = 30$
Minimum value of $f(-1, 2, -5) = -30$

127. $\left(\pm\frac{4}{3}, -\frac{4}{3}, -\frac{4}{3}\right)$ 128. $x + xy, x + xy + \frac{1}{2}xy^2$

129. xy, xy

130. $y + \frac{1}{2}(2xy - y^2), y + \frac{1}{2}(2xy - y^2) + \frac{1}{6}(3x^2y - 3xy^2 + 2y^3)$

7 Applications of Partial Differentiation

Curvature measures how much a curve deviates from a flat surface. It is applied to both one-dimensional and two-dimensional space. In two-dimensional space, curvature measures how rapidly the curve bends at a given point.

Curvature plays an important role in fields like general relativity, the design of roads or aircraft wings, etc. First, we will see the curvature of any curve at a point.

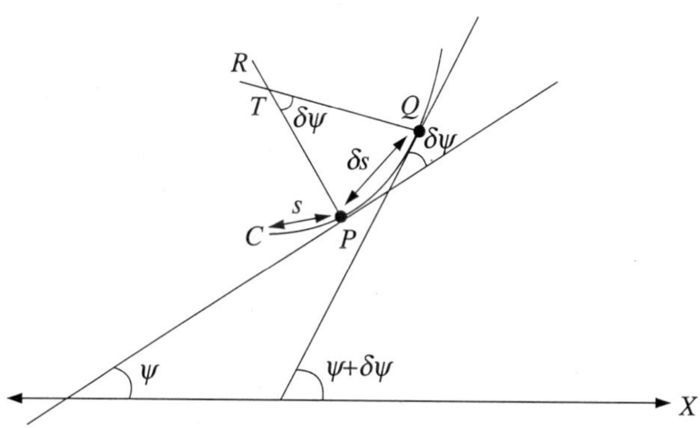

FIGURE 7.1 Curvature.

Let's take two points on the curve say P and Q. Tangents at points P and Q made an angle ψ and $\psi + \delta\psi$ respectively with the x-axis. Normal at both the points will intersect at the point S. Suppose the length of arc CP and PQ is p and δs respectively.

Now, the curvature of the curve at point P is possible when $Q \to P$, $\delta s \to 0$ and $\delta\psi \to 0$.

Where $\dfrac{\delta\psi}{\delta s}$ is the average curvature of arc PQ.

$$\lim_{\delta s \to 0} \frac{\delta\psi}{\delta s} = \frac{d\psi}{ds} = (\text{say}\,\kappa)$$

Means κ is the curvature of a curve at point P.

7.1 CURVATURE OF A CIRCLE

The curvature of a circle measures how a circle deviates from being a straight line. The circle that best approaches a given curve at a point is called the ***circle of curvature***. The center and radius of the circle of curvature are called the ***center of curvature*** and ***radius of curvature*** respectively.

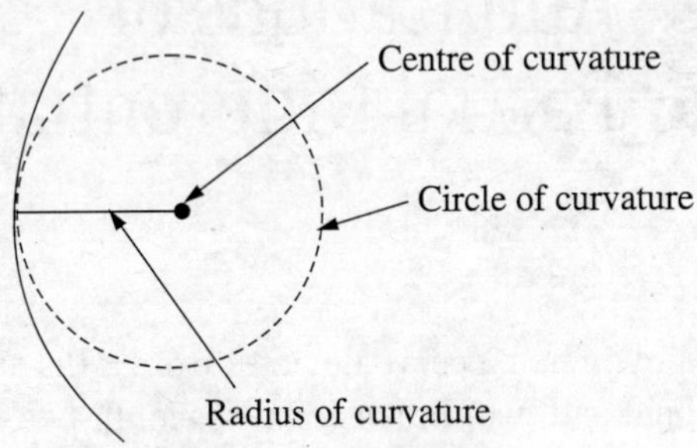

FIGURE 7.2 Curvature of circle.

For a circle the curvature κ proportional to the reciprocal of the radius r, i.e., $\rho = \frac{1}{\kappa} = \frac{ds}{d\psi}$. That means the radius of the circle is larger then, the radius of curvature will be small and it seems a very curvy curve. In the reverse case, when the curvature κ is small then the radius ρ becomes large, the curve becomes also straight. One can say that straight lines have zero curvature.

EXAMPLE 7.1 Find the radius of curvature of the curve $S = 9a\sin^2\left(\frac{\psi}{3}\right)$.

Ans. We have, $\rho = \frac{ds}{d\psi} = 9a.2\sin\left(\frac{\psi}{3}\right)\cdot\cos\left(\frac{\psi}{3}\right)\cdot\frac{1}{3}$

$$\therefore \quad \rho = 3a.\sin 2\left(\frac{\psi}{3}\right)$$

$$\therefore \quad \rho = 3a.\sin\left(\frac{2\psi}{3}\right)$$

7.1.1 Radius of Curvature in Cartesian Form

Radius of curvature measures how sharply a curve bends at a given point. For deriving Cartesian form, assume a point $P(x, y)$ on a curve $y = f(x)$.

Now, $\tan\psi = \frac{dy}{dx}$.

Differentiating this with respect to S,

$$\sec^2\psi \frac{d\psi}{dS} = \frac{d^2y}{dx^2}\frac{dx}{dS}$$

Here, $y = f(x)$, $\rho = f(r)$, If $\tan\psi = \frac{dy}{dx}$ then $\sin\psi = \frac{dy}{dS}$ and $\cos\psi = \frac{dx}{dS}$.

$$\therefore \qquad \sec^2\psi \frac{1}{\rho} = \frac{d^2y}{dx^2}\cos\psi$$

$$\therefore \qquad \sec^2\psi \frac{1}{\rho} = \frac{d^2y}{dx^2}\frac{1}{\sec\psi}$$

$$\therefore \qquad \frac{\sec^3\psi}{d^2y/dx^2} = \rho$$

Let's do some manipulation, $\therefore \rho = \frac{(\sec^2\psi)^{\frac{3}{2}}}{d^2y/dx^2} = \frac{(1+\tan^2\psi)^{\frac{3}{2}}}{d^2y/dx^2}$

$$\therefore \qquad \rho = \frac{\left(1+\left(\frac{dy}{dx}\right)^2\right)^{\frac{3}{2}}}{\frac{d^2y}{dx^2}}$$

EXAMPLE 7.2 Find the radius of curvature for given Cartesian curves: (i) $a^2y = x^3 - a^3$; (ii) $y^2 = 4ax$.

Ans. (i) Given $a^2y = x^3 - a^3$

Differentiating with respect to x, $a^2\frac{dy}{dx} = 3x^2$

$$\frac{dy}{dx} = \frac{3x^2}{a^2} \Rightarrow \frac{d^2y}{dx^2} = \frac{6x}{a^2}$$

Hence, $\rho = \frac{\left(1+\left(\frac{dy}{dx}\right)^2\right)^{\frac{3}{2}}}{\frac{d^2y}{dx^2}} = \frac{\left(1+\left(\frac{3x^2}{a^2}\right)^2\right)^{\frac{3}{2}}}{\frac{6x}{a^2}}$

$$\therefore \qquad \rho = \frac{\left(1+\frac{9x^4}{a^4}\right)^{\frac{3}{2}}}{\frac{6x}{a^2}} = \frac{(a^4+9x^4)}{6xa^4}$$

(ii) Given $y^2 = 4ax$

Differentiating with respect to x, $2y\dfrac{dy}{dx} = 4a$

$$\frac{dy}{dx} = \frac{2a}{y}$$

$$\Rightarrow \frac{d^2y}{dx^2} = -\frac{2a}{y^2}\frac{dy}{dx} = -\frac{2a}{y^2}\left(\frac{2a}{y}\right) = -\frac{4a^2}{y^3}$$

Now,
$$\rho = \frac{\left(1+\left(\dfrac{2a}{y}\right)^2\right)^{\frac{3}{2}}}{-\dfrac{4a^2}{y^3}} = \frac{(y^2+4a^2)^{\frac{3}{2}}}{-4a^2}$$

$\therefore$
$$\rho = \frac{(4ax+4a^2)^{\frac{3}{2}}}{-4a^2} = \frac{(4a)^{3/2}(x+a)^{3/2}}{-4a^2}$$

Hence,
$$\rho = -\frac{2(x+a)^{3/2}}{-a^{1/2}}$$

But length cannot be negative, $\Rightarrow \rho = \dfrac{2(x+a)^{3/2}}{-a^{1/2}}$.

7.1.2 Parametric Form

As we have assumed a curve $y = f(x)$ which should be a twice differentiable function. Consider a point $P(x, y)$ on a curve $y = f(x)$. Then the parametric form of this point is defined by, $x = x(t)$, $y = y(t)$.

$$\frac{dy}{dx} = \frac{dy/dt}{dx/dt}$$

$\therefore$
$$\frac{d^2y}{dx^2} = \frac{d}{dt}\left(\frac{dy/dt}{dx/dt}\right)\frac{dt}{dx} = \frac{\dfrac{dx}{dt}\cdot\dfrac{d^2y}{dt^2} - \dfrac{dy}{dt}\cdot\dfrac{d^2x}{dt^2}}{\left(\dfrac{dx}{dt}\right)^2}\cdot\frac{1}{\dfrac{dx}{dt}}$$

$\therefore$
$$\frac{d^2y}{dx^2} = \frac{\dfrac{dx}{dt}\cdot\dfrac{d^2y}{dt^2} - \dfrac{dy}{dt}\cdot\dfrac{d^2x}{dt^2}}{\left(\dfrac{dx}{dt}\right)^3}$$

Now, $$\rho = \frac{\left(1+\left(\frac{dy}{dx}\right)^2\right)^{\frac{3}{2}}}{\frac{d^2y}{dx^2}} = \frac{\left(1+\left(\frac{dy/dt}{dx/dt}\right)^2\right)^{\frac{3}{2}}}{\left(\frac{dx}{dt}\cdot\frac{d^2y}{dt^2}-\frac{dy}{dt}\cdot\frac{d^2x}{dt^2}\right)\Big/\left(\frac{dx}{dt}\right)^3}$$

$\therefore$ $$\rho = \frac{\left(\left(\frac{dx}{dt}\right)^2+\left(\frac{dy}{dt}\right)^2\right)^{\frac{3}{2}}}{\left(\frac{dx}{dt}\cdot\frac{d^2y}{dt^2}-\frac{dy}{dt}\cdot\frac{d^2x}{dt^2}\right)}.$$

EXAMPLE 7.3 Find the radius of curvature for given polar curves:

(i) $x = a\cos^3 t,\ y = a\sin^3 t$

(ii) $x = a(\theta + \sin\theta),\ y = a(1 - \cos\theta)$

(iii) $r^m = a^m \cos m\theta$

Ans. (i) $x = a\cos^3 t,\ y = a\sin^3 t$

$$\frac{dx}{dt} = -3a\cos^2 t\sin t,\ \frac{dy}{dt} = 3a\sin^2 t\cos t$$

$$\frac{dy}{dx} = \frac{dy/dt}{dx/dt} = \frac{-3a\cos^2 t\sin t}{3a\sin^2 t\cos t} = -\tan t$$

$$\frac{d^2y}{dx^2} = -\sec^2 t\frac{dt}{dx} = -\sec^2 t\left(\frac{1}{-3a\cos^2 t\sin t}\right) = \frac{1}{3a}\sec^4 t\,.\,\text{cosec}\ t$$

Now, $$\rho = \frac{\left(1+\left(\frac{dy}{dx}\right)^2\right)^{\frac{3}{2}}}{\frac{d^2y}{dx^2}} = \frac{(1+\tan^2 t)^{\frac{3}{2}}}{\frac{1}{3a}\sec^4 t\,.\,\text{cosec}\ t} = 3a\cos t\sin t$$

(ii) $x = a(\theta + \sin\theta),\ y = a(1 - \cos\theta)$

$$\frac{dx}{d\theta} = a(1+\cos\theta),\ \frac{dy}{d\theta} = a\sin\theta$$

$$\frac{d^2x}{d\theta^2} = -a\sin\theta,\ \frac{d^2y}{d\theta^2} = a\cos\theta$$

$$\text{Now,}\quad \rho = \frac{\left(\left(\frac{dx}{dt}\right)^2 + \left(\frac{dy}{dt}\right)^2\right)^{\frac{3}{2}}}{\left(\frac{dx}{dt}\cdot\frac{d^2y}{dt^2} - \frac{dy}{dt}\cdot\frac{d^2x}{dt^2}\right)} = \frac{\left((a(1+\cos\theta))^2 + (a\sin\theta)^2\right)^{\frac{3}{2}}}{(a(1+\cos\theta)\cdot(a\cos\theta) - a\sin\theta\cdot(-a\sin\theta))}$$

$$\therefore\quad \rho = \frac{(4a^2(1+\cos\theta)^2)^{3/2}}{a^2(1+\cos\theta)} = 8a(1+\cos\theta)^2$$

(iii) $r^m = a^m \cos m\theta$

Taking log on both sides,

$\therefore\ m\log r = m\log a + \log\cos m\theta$

Differentiating w.r.t. θ we have,

$$\frac{m}{r}r' = -m\tan m\theta$$

$$\therefore\ r' = -r\tan m\theta$$

Differentiating w.r.t. θ we have,

$$r'' = -(r'\tan m\theta + rm\sec^2 m\theta)$$

$$\therefore\ r'' = -((-r\tan m\theta)\tan m\theta + rm\sec^2 m\theta)\quad \text{(substituting } r')$$

$$\therefore\ r'' = r\tan^2 m\theta - rm\sec^2 m\theta$$

$$\text{Now,}\ \rho = \frac{(r^2 + r^2\tan^2 m\theta)^{3/2}}{r^2 + 2r^2\tan^2 m\theta - r\tan^2 m\theta + r^2 m\sec^2 m\theta} = \frac{r}{m+1}\sec m\theta$$

7.1.3 Radius of Curvature of Pedal Curves

As we have already studied the pedal equation for the given curve in Chapter 5, with that reference, we will understand how to derive the radius of curvature of the pedal curve.

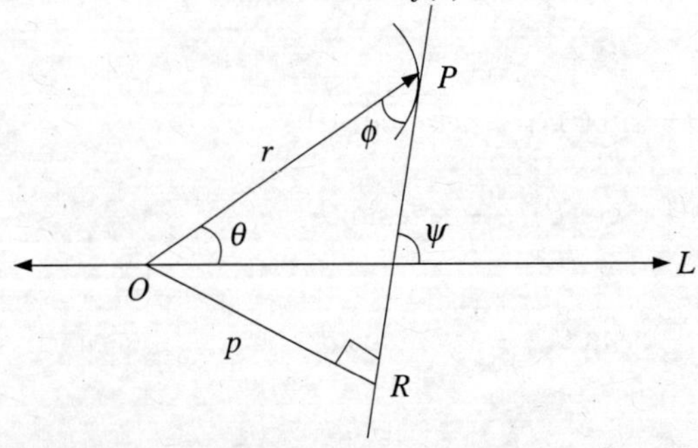

FIGURE 7.3 Radius of curvature of Pedal Curves.

We have,

$$p = r\sin\phi$$

Differentiate with respect to r we have,

$$\frac{dp}{dr} = r\cos\phi\,\frac{d\phi}{dr} + \sin\phi$$

$$\therefore \quad \frac{dp}{dr} = r\,\frac{dr}{ds}\,\frac{d\phi}{dr} + r\,\frac{d\theta}{ds}$$

$$\therefore \quad \frac{dp}{dr} = r\frac{d\phi}{ds} + r\frac{d\theta}{ds} + r\left(\frac{d\phi}{ds} + \frac{d\theta}{ds}\right)$$

Also, we have $\psi = \theta + \phi$,

$$\frac{d\psi}{ds} = \frac{d\theta}{ds} + \frac{d\phi}{ds}$$

$$\therefore \quad \frac{dp}{dr} = r\frac{d\psi}{ds}$$

We have, $\rho = \dfrac{ds}{d\psi}$

$$\therefore \quad \frac{dp}{dr} = r\frac{1}{\rho} \Rightarrow \boxed{\rho = r\frac{dr}{dp}}$$

EXAMPLE 7.4 Find the radius of curvature of $r^3 = 2ap^2$.

Ans. Here, $r^3 = 2ap^2$

Differentiating w.r.t. p,

$$3r^2\frac{dr}{dp} = 4ap$$

$$\frac{dr}{dp} = \frac{4ap}{3r^2}$$

Now, $\rho = r\dfrac{dr}{dp} = r\,.\,\dfrac{4ap}{3r^2} = \dfrac{4ap}{3r}$

(From the given equation, $\sqrt{\dfrac{r^3}{2a}} = p$)

$$\therefore \quad \rho = \frac{4ap}{3r} = \frac{4a\left(\dfrac{r^3}{2a}\right)^{1/2}}{3r} = \frac{2}{3}\sqrt{2ar}$$

EXAMPLE 7.5 Find the radius of curvature of $p^2 = \dfrac{r^4}{r^2 + a^2}$.

Ans. Given, $p^2 = \dfrac{r^4}{r^2 + a^2}$

$$\therefore \quad \frac{1}{p^2} = \frac{r^2 + a^2}{r^4} = \frac{1}{r^2} + \frac{a^2}{r^4}$$

Differentiating w.r.t. r,

$$\frac{-2}{p^3}\frac{dp}{dr} = \frac{-2}{r^3} - \frac{4a^2}{r^5}$$

$$\therefore \quad \frac{dp}{dr} = p^3\left(\frac{1}{r^3} + \frac{2a^2}{r^5}\right)$$

$$\therefore \quad \frac{dp}{dr} = p^3\left(\frac{1}{r^3} + \frac{2a^2}{r^5}\right)$$

$$\therefore \quad \frac{dp}{dr} = (p^2)^{3/2}\left(\frac{r^2 + 2a^2}{r^5}\right)$$

$$\therefore \quad \frac{dp}{dr} = \left(\frac{r^4}{r^2 + a^2}\right)^{3/2}\left(\frac{r^2 + 2a^2}{r^5}\right)$$

$$\therefore \quad \frac{dp}{dr} = \frac{r^6}{(r^2 + a^2)^{3/2}}\left(\frac{r^2 + 2a^2}{r^5}\right)$$

$$\therefore \quad \frac{dp}{dr} = \frac{r(r^2 + 2a^2)}{(r^2 + a^2)^{3/2}}$$

Hence, $r\dfrac{dr}{dp} = \rho = \dfrac{(r^2 + a^2)^{3/2}}{(r^2 + 2a^2)}$

7.1.4 Tangential Polar Form

The radius of curvature in the tangential polar form can be derived from the general formula for the radius of curvature in polar coordinates. Let's go through the steps systematically.

$$\rho = \frac{ds}{d\psi} = r\frac{dr}{dp}$$

$$\frac{dr}{ds} = \cos\phi \text{ and } p = r\sin\phi$$

$$\frac{dp}{d\psi} = \frac{dp}{dr}\frac{dr}{ds}\frac{ds}{d\psi} = \frac{dp}{dr}.\cos\phi\,.\,r\frac{dr}{dp}$$

$$\therefore \qquad \frac{dp}{d\psi} = r\cos\phi$$

Now, $(r\cos\phi)^2 + (r\sin\phi)^2 = \left(\frac{dp}{d\psi}\right)^2 + p^2$

$$\therefore \qquad r^2 = p^2 + \left(\frac{dp}{d\psi}\right)^2$$

Differentiating w.r.t. p

$$\therefore \qquad 2r\frac{dr}{dp} = 2p + 2\left(\frac{dp}{d\psi}\right)\frac{d}{d\psi}\left(\frac{dp}{d\psi}\right)\frac{d\psi}{dp}$$

$$\therefore \qquad r\frac{dr}{dp} = p + \left(\frac{d^2p}{d\psi^2}\right)$$

Hence, $\rho = p + \left(\frac{d^2p}{d\psi^2}\right)$

EXAMPLE 7.6 Find the radius of curvature of the curve $p = a\sin b\psi$.

Ans. Given, $p = a\sin b\psi$

Hence, $\quad \frac{dp}{d\psi} = ab\cos b\psi$ and $\frac{d^2p}{d\psi^2} = -ab^2\sin b\psi$

Now, $\quad \rho = p + \left(\frac{d^2p}{d\psi^2}\right) = a\sin b\psi - ab^2\sin b\psi$

$$\therefore \qquad \rho = a\sin b\psi(1-b^2)$$

$$\therefore \qquad \rho = p(1-b^2)$$

7.2 RADIUS OF CURVATURE AT THE ORIGIN

7.2.1 Substitution Method

The radius of curvature at the origin means we find the curvature by replacing (x, y) by $(0, 0)$. As we know the formula, $\rho_{(x,y)} = \dfrac{\left(1+\left(\frac{dy}{dx}\right)^2\right)^{\frac{3}{2}}}{\frac{d^2y}{dx^2}}$.

$$\text{At origin, } \rho_{(0,0)} = \frac{\left(1+\left(\frac{dy}{dx}\right)^2_{(0,0)}\right)^{\frac{3}{2}}}{\left(\frac{d^2y}{dx^2}\right)_{(0,0)}}$$

7.2.2 Expanded Method

Continue with the initially assumed curve $y = f(x)$ which passes through the origin $(x, y) = (0\ ,0)$. Hence, $y_0 = f(0) = 0$.

Suppose, y can be expanded in powers of y, i.e., $y = px + q\frac{x^2}{2!} + ...$

By Maclaurin series theorem expansion will be,

$$y = (y)_{(0,0)} + x(y_1)_{(0,0)} + \frac{x^2}{2!}(y_2)_{(0,0)} + \frac{x^3}{3!}(y_3)_{(0,0)} \cdots$$

$$y = x(y_1)_{(0,0)} + \frac{x^2}{2!}(y_2)_{(0,0)} + ...$$

From the above two expressions,

$$px + q\frac{x^2}{2!} + ... = x(y_1)_{(0,0)} + \frac{x^2}{2!}(y_2)_{(0,0)} + ...$$

$$(y_1)_{(0,0)} = \left.\frac{dy}{dx}\right|_{(0,0)} = p \text{ and } (y_2)_{(0,0)} = \left.\frac{d^2y}{dx^2}\right|_{(0,0)} = q$$

$$\therefore \qquad \rho_{(0,0)} = \frac{(1+p^2)^{\frac{3}{2}}}{q}$$

EXAMPLE 7.7 Find the radius of curvature of the curve $a(y^2 - x^2) = x^3)$.

Ans. Given $a(y^2 - x^2) = x^3)$

$$\therefore \qquad ay^2 - ax^2 = x^3$$

$$\therefore \qquad y = \pm x\left(\frac{x+a}{a}\right)^{1/2} = \pm x\left(1+\frac{x}{a}\right)^{1/2}$$

$$\therefore \qquad y = \pm x\left(1+\frac{1}{2}\frac{x}{a}+...\right) = \pm\left(x+\frac{1}{2}\frac{x^2}{a}\right)$$

Comparing to the equation, $p = \pm 1$ and $q = \pm\frac{1}{a}$

Hence, $\rho_{(0,0)} = \dfrac{(1+p^2)^{\frac{3}{2}}}{q} = \dfrac{(1+1^2)^{\frac{3}{2}}}{\pm(1/a)}$

Now, calculating for both the values,

$$\rho_{(0,0)} = \frac{(1+p^2)^{\frac{3}{2}}}{q} = \frac{(1+1^2)^{\frac{3}{2}}}{(1/a)} = \sqrt{8}a \text{ and}$$

$$\rho_{(0,0)} = \frac{(1+p^2)^{\frac{3}{2}}}{q} = \frac{(1+(-1)^2)^{\frac{3}{2}}}{-(1/a)} = \sqrt{8}a \text{ (radius can't be negative)}$$

7.2.3 Newton's Method

This method is only applicable when the curve passes through the origin and therefore x-axis or y-axis will be tangent there. Consider when the x-axis is tangent then,

By Maclaurin series theorem expansion will be,

$$y = (y)_{(0,0)} + x(y_1)_{(0,0)} + \frac{x^2}{2!}(y_2)_{(0,0)} + \frac{x^3}{3!}(y_3)_{(0,0)} \cdots$$

At point $(x, y) = (0,0)$, $\left.\dfrac{dy}{dx}\right|_{(0,0)} = y_1\big|_{(0,0)} = p = 0$,

$$\therefore \qquad y = \frac{x^2}{2!}(y_2)_{(0,0)} + \frac{x^3}{3!}(y_3)_{(0,0)} \cdots$$

Dividing by x^2 and taking limit on both sides it will convert,

$$\lim_{x\to 0}\frac{y}{x^2} = \lim_{x\to 0}\left(\frac{1}{2!}(y_2)_{(0,0)} + \frac{x}{3!}(y_3)_{(0,0)} \cdots\right)$$

$$\therefore \qquad \lim_{x\to 0}\frac{y}{x^2} = \lim_{x\to 0}\left(\frac{1}{2!}(y_2)_{(0,0)} + 0 + \ldots\right)$$

$$\therefore \qquad \lim_{x\to 0}\frac{2y}{x^2} = \lim_{x\to 0}((y_2)_{(0,0)}) = q$$

Now, $\qquad \rho_{(0,0)} = \dfrac{(1+p^2)^{\frac{3}{2}}}{q} = \dfrac{1}{q}$

$$\therefore \qquad \rho_{(0,0)} = \lim_{x \to 0} \frac{x^2}{2y}$$

Similarly, when the curve passes through the origin and the y-axis is tangent then, $\rho_{(0,0)} = \lim_{y \to 0} \frac{y^2}{2x}$.

Note: When the curve passes through the origin and is expressed in algebraic equation form, then tangent at the origin can obtained by equating a lower degree term to zero.

EXAMPLE 7.8 Find the radius of curvature at the origin of the curve $x^3y - xy^3 + 2x^2y + xy - y^2 + 2x = 0$.

Ans. Equating lower degree term to zero, $x = 0$.
Therefore, the tangent is y-axis.

Hence, $\rho_{(0,0)} = \lim_{y \to 0} \frac{y^2}{2x}$

Dividing the given equation by $2x$ we have,

$$\frac{x^3y - xy^3 + 2x^2y + xy - y^2 + 2x}{2x} = 0$$

$$x^3 \frac{y}{2x} - xy\frac{y^2}{2x} + 2x^2 \frac{y}{2x} + x\frac{y}{2x} - \frac{y^2}{2x} + 1 = 0$$

Taking limit $y \to 0$ we have, $\rho = 1$.

7.3 CENTRE OF CURVATURE AND CIRCLE OF CURVATURE

Let $P(x, y)$ be a point on the curve and $R(\alpha, \beta)$ be the center of curvature on the given circle of curvature. Take a tangent at point $P(x, y)$ cuts on the x-axis at an angle ψ and PR be a radius of curvature. We need to find the values of α and β.

$$\tan\psi = \frac{dy}{dx} = y_1$$

$$\sin\psi = \frac{y_1}{\sqrt{1+y_1^2}} \text{ and } \cos\psi = \frac{1}{\sqrt{1+y_1^2}}$$

We know the formula of the radius of curvature, $\rho = \frac{(1+y_1^2)^{3/2}}{y_2}$.

From Figure 7.2, consider ΔPQR, $\sin\psi = \frac{QP}{PR} = \frac{QP}{\rho} \Rightarrow QP = \rho\sin\psi$

Similarly, $QR = \rho\cos\psi$.

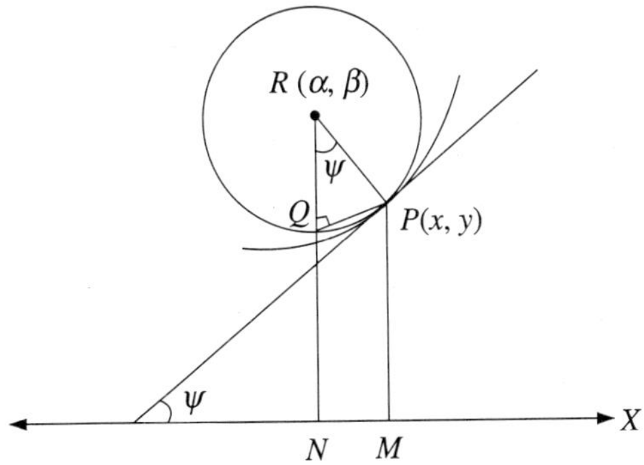

FIGURE 7.4 Centre and Circle of Curvature.

Now, the distance between the origin and to point Q is α and to point P is x then,

$$PQ = x - \alpha = \rho \sin \psi$$

$$\therefore \qquad \alpha = x - \rho \sin \psi = x - \frac{(1+y_1^2)^{3/2}}{y_2} \cdot \frac{y_1}{\sqrt{1+y_1^2}}$$

$$\therefore \qquad \alpha = x - \frac{(1+y_1^2)y_1}{y_2}$$

In a similar way,

$$QR = \beta - y = \rho \cos \psi$$

$$\therefore \qquad \beta = y + \frac{1+y_1^2}{y_2}$$

Thus, the center will be $R(\alpha, \beta) = \left(x - \dfrac{(1+y_1^2)y_1}{y_2}, y + \dfrac{1+y_1^2}{y_2}\right)$ and the equation of the circle of curvature is: $(x-\alpha)^2 + (y-\beta)^2 = \rho^2$.

EXAMPLE 7.9 Find the circle of curvature of the curve $xy = c^2$ at point (c, c).

Ans. Here, $xy = c^2$

Then, $$y + x\frac{dy}{dx} = 0 \Rightarrow \frac{dy}{dx} = -\frac{y}{x}$$

At point (c, c), $\dfrac{dy}{dx} = -\dfrac{c}{c} = -1$

and $$\frac{d^2y}{dx^2} = -\frac{x\dfrac{dy}{dx} - y}{x^2}$$

At point (c, c), $\dfrac{d^2y}{dx^2} = \dfrac{2}{c}$

Now, $$\alpha = x - \frac{(1+y_1^2)y_1}{y_2} = c - \frac{(1+1^2)(-1)}{2/c} = 2c$$

and $$\beta = y + \frac{1+y_1^2}{y_2} = c + \frac{1+1^2}{2/c} = 2c$$

and $$\rho = \frac{(1+y_1^2)^{3/2}}{y_2} = \frac{(1+1^2)^{3/2}}{2/c} = c\sqrt{2}$$

Finally, $(x-\alpha)^2 + (y-\beta)^2 = \rho^2 \Rightarrow (x-2c)^2 + (y-2c)^2 = 2c$

EXAMPLE 7.10 Find the circle of curvature of the curve $x = a \log \tan\left(\dfrac{\pi}{4} + \dfrac{\theta}{2}\right)$ and $y = a \sec \theta$ at any point.

Ans. Given $x = a \log \tan\left(\dfrac{\pi}{4} + \dfrac{\theta}{2}\right)$

$$\therefore \quad \frac{dx}{d\theta} = \frac{a}{\tan\left(\frac{\pi}{4} + \frac{\theta}{2}\right)} \cdot \sec^2\left(\frac{\pi}{4} + \frac{\theta}{2}\right) \cdot \frac{1}{2}$$

$$\therefore \quad \frac{dx}{d\theta} = \frac{a}{\sin 2\left(\left(\frac{\pi}{4} + \frac{\theta}{2}\right)\right)} = \frac{a}{\sin\left(\left(\frac{\pi}{2} + \theta\right)\right)} = a \sec \theta$$

and $$y = a \sec \theta \Rightarrow \frac{dy}{d\theta} = a \sec \theta \tan \theta$$

Hence, $$\frac{dy}{dx} = \frac{a \sec \theta \tan \theta}{a \sec \theta} = \tan \theta$$

$$\therefore \quad \frac{d^2y}{dx^2} = \sec^2\theta \frac{d\theta}{dx} = \sec^2\theta \cdot \frac{1}{a \sec \theta} = \frac{\sec \theta}{a}$$

Now, $$\alpha = x - \frac{(1+y_1^2)y_1}{y_2} = a \log \tan\left(\frac{\pi}{4} + \frac{\theta}{2}\right) - \frac{(1+\tan^2\theta)\tan\theta}{\sec \theta / a}$$

$$\therefore \quad \alpha = a \log \tan\left(\frac{\pi}{4} + \frac{\theta}{2}\right) - a \sec \theta \tan \theta$$

then $$\beta = y + \frac{1+y_1^2}{y_2} = a\sec\theta + \frac{1+\tan^2\theta}{\sec\theta/a} = 2a\sec\theta$$

and $$\rho = \frac{(1+y_1^2)^{3/2}}{y_2} = \frac{(1+\tan^2\theta)^{3/2}}{\sec\theta/a} = a\sec^2\theta$$

Equation of circle of curvature, $(x-\alpha)^2 + (y-\beta)^2 = \rho^2$

$$\therefore \left(x - \left(a\log\tan\left(\frac{\pi}{4}+\frac{\theta}{2}\right) - a\sec\theta\tan\theta\right)\right)^2 + (y - 2a\sec\theta)^2 = a^2\sec^4\theta$$

7.4 CHORD OF CURVATURE

A chord of curvature refers to a specific geometric property related to a curve at a given point. It is the straight-line segment that connects two points on the curve.

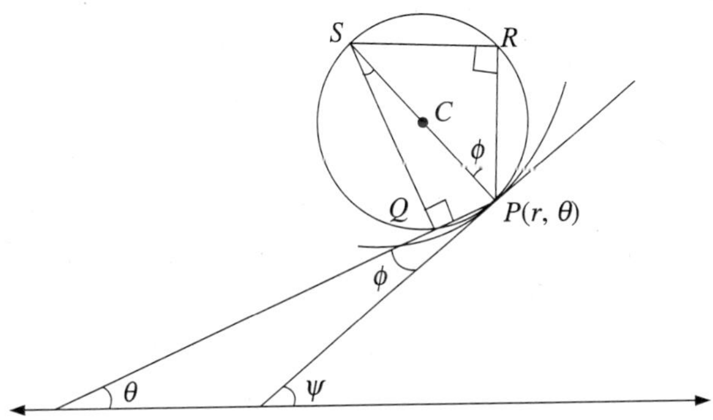

FIGURE 7.5 Chord of Curvature.

A chord drawn through point P of a circle of curvature is called a chord of curvature at point P.

Note that, $\angle SPR = \phi$, $\angle SPQ = \frac{\pi}{2} - \phi$, $SP = 2CP = 2\rho$

(i) **Chord of curvature through the origin at point P:**

Consider, $\cos\left(\frac{\pi}{2} - \phi\right) = \frac{PQ}{SP}$

Hence, $PQ = SP\cos\left(\frac{\pi}{2} - \phi\right) = 2\rho\sin\phi$

Chord of curvature through the origin, $2\rho\sin\phi$.

(ii) **Chord of curvature perpendicular to radius vector at *P*:**

Consider, $\cos\phi = \dfrac{PR}{SP} = \dfrac{PR}{2\rho}$

Hence, the Chord of curvature perpendicular to the radius vector, $PR = 2\rho\cos\phi$.

For Cartesian curves:

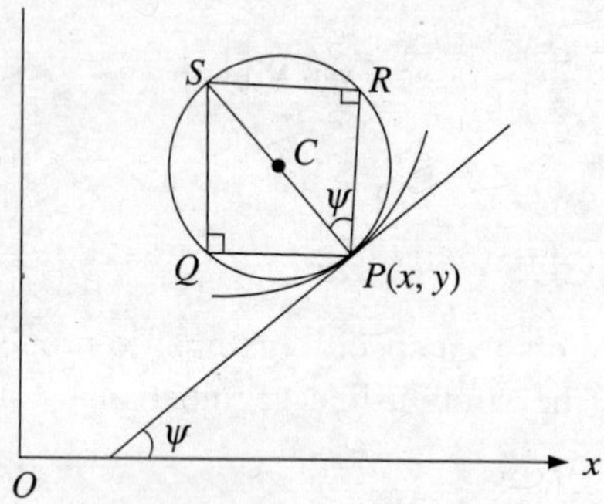

FIGURE 7.6 Chord of curvature in the Cartesian curve.

Parallel to the x-axis at *P*, $\sin\psi = \dfrac{PQ}{SP} = \dfrac{PQ}{2\rho}$

Hence, the chord of curvature parallel to the *x*-axis, $PQ = 2\rho\sin\psi$.

Similarly, the chord of curvature parallel to the *y*-axis, $PR = 2\rho\cos\psi$.

EXAMPLE 7.11 Find the chord of curvature through the pole $r = a(1 + \cos\theta)$.

Ans. $r = a(1 + \cos\theta) \Rightarrow \dfrac{dr}{d\theta} = -a\sin\theta$

$$\tan\phi = r\frac{d\theta}{dr} = \frac{a(1+\cos\theta)}{-a\sin\theta} = -\frac{1 + 2\cos^2(\theta/2) - 1}{2\sin(\theta/2)\cos(\theta/2)} = -\cot(\theta/2) = \tan(\pi/2 + \theta/2)$$

$\therefore\ \phi = (\pi/2 + \theta/2)$

Now, $p = r\sin\phi = r\sin\left(\dfrac{\pi}{2} + \dfrac{\theta}{2}\right) = r\cos\dfrac{\theta}{2}$

Squaring on both sides and multiplying by 2,

$$\therefore \qquad 2p^2 = 2r^2\cos^2\frac{\theta}{2} = 2r^2\frac{(1+\cos\theta)}{2} = \frac{r^3}{a}$$

$$\left(\therefore p = \frac{r^{3/2}}{\sqrt{2a}}\right)$$

Differentiating with respect to r,

$$\therefore \qquad 4p\frac{dp}{dr} = \frac{3r^2}{a} \Rightarrow \frac{dp}{dr} = \frac{3r^2}{a.4p}$$

$$\text{Now,} \qquad \rho = r\frac{dr}{dp} = r\frac{4ap}{3r^2} = \frac{4ap}{3r}$$

$$\therefore \qquad \rho = \frac{4a}{3r}\left(\frac{r^{3/2}}{\sqrt{2a}}\right) = \frac{2\sqrt{2ar}}{3}$$

$$\text{Chord of curvature} = 2\rho\sin\phi = 2\left(\frac{2\sqrt{2ar}}{3}\right)\sin\left(\frac{\pi}{2}+\frac{\theta}{2}\right) = \frac{4}{3}\sqrt{2ar}\cos\frac{\theta}{2}$$

$$= \frac{4}{3}\sqrt{2ar}\sqrt{\frac{r}{2a}} = \frac{4}{3}r$$

EXAMPLE 7.12 Prove that the chord of curvature through the origin of the curve $p = f(r)$ is $\dfrac{2f(r)}{f'(r)}$.

Ans. Given, $p = f(r) = r\sin\phi$ and $\rho = r\dfrac{dr}{dp}$

$$p = f(r) \Rightarrow \frac{dp}{dr} = f'(r).$$

Hence, $\rho = r\dfrac{1}{f'(r)}$

Chord of curvature, $2\rho\sin\phi = \dfrac{2r}{f'(r)}\cdot\dfrac{f(r)}{r} = \dfrac{2f(r)}{f'(r)}$

7.5 ASYMPTOTES

An asymptote is an imaginary line that is very close to the whole graph of a function. A curve's asymptote is the point at which the curve converges but, the curve and its asymptotes approach each other infinitely and never meet. There are two types of asymptotes:

1. **Rectangular Asymptote:** If an asymptote is parallel to the x-axis or y-axis then it is called rectangular asymptote. An asymptote that is parallel to the x-axis is called a horizontal asymptote and parallel to the y-axis is called a vertical asymptote.

   Horizontal asymptotes are smooth flat lines parallel to the x-axis and graph of a function $y = f(x)$ approaches to constant value and

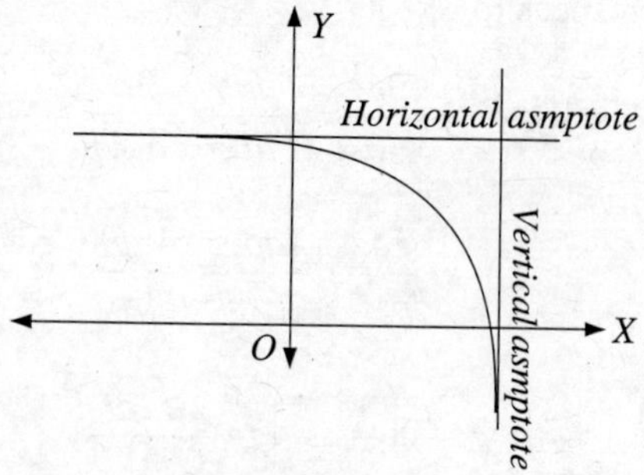

FIGURE 7.7

x goes to $\pm\infty$. In vertical asymptotes, the curve goes to $\pm\infty$ and x approaches to constant.

Method to find asymptote parallel to co-ordinate axis:

(a) **Parallel to x-axis:** Equate the highest coefficient of power of x to zero in the equation of the curve.

(b) **Parallel to the y-axis:** Equate the highest coefficient of power of y to zero in the equation of the curve.

2. **Oblique Asymptote:** An asymptote that is neither parallel to the x-axis or the y-axis is treated as an oblique asymptote. It occurs when x goes to $\pm\infty$, the curve moves towards a line $y = mx + c$. Oblique asymptotes arise when the degree of the polynomial in the numerator is more than that polynomial in the denominator.

Method: Let the equation of the curve be,

$$\phi_n(x, y) + \phi_{n-1}(x, y) + \ldots + \phi_1(x, y) + k = 0$$

Step 1: Find $\phi_n(m)$ by putting $x = 1$ and $y = m$ in the equation.

Step 2: Solve $\phi_n(m) = 0$ and find all real roots.

Step 3: If m is non-repeated root then $c = -\dfrac{\phi_{n-1}(m)}{\phi_n'(m)}$ $(\phi_n'(m) \neq 0)$.

Here, if $\phi'_n(m) = 0$ then there is no asymptote corresponding to the value m.

Step 4: If m is repeated twice then c has two values given by,

$$\frac{c^2}{2!}\phi_n''(m) + \frac{c}{1!}\phi_{n-1}'(m) + \phi_{n-2}(m) = 0 \quad (\phi_n''(m) \neq 0)$$

If m is repeated thrice then c has two values given by,

$$\frac{c^3}{3!}\phi_4''(m) + \frac{c^2}{2!}\phi_3''(m) + \frac{c}{1!}\phi_2'(m) + \phi_1(m) = 0 \quad (\phi_n''(m) \neq 0)$$

Finally, an asymptote of the curve, $y = mx + c$

EXAMPLE 7.13 Find the asymptotes of the curve $yx^2 - 4x^2 + 2y + 2 = 0$.

Ans. Since the highest degree terms are in x, the curve is parallel to the x-axis. Now equate it to zero,

$$x^2(y-4) = 0$$

$(y - 4) = 0 \Rightarrow y = 4$ is an asymptote.

EXAMPLE 7.14 Find the asymptotes of the curve $y^3x - 2y^3 + y^2 + x^2 + 2 = 0$.

Ans. Since the highest degree terms are in y, the curve is parallel to the y-axis.

Equate it to zero, $y^3(x-2) = 0 \Rightarrow x = 2$ is an asymptote.

EXAMPLE 7.15 Find the asymptotes of the curve $(x^2 - y^2)(x + 2y) + 5(x^2 + y^2) + x + y = 0$.

Ans. Here, a rectangular asymptote does not exist. We have to find an oblique asymptote,

$$\phi_3(m) = (x^2 - y^2)(x + 2y)$$

$$\phi_2(m) = 5(x^2 + y^2)$$

$$\phi_1(m) = x + y$$

Put $x = 1$ and $y = m$,

$$\phi_3(m) = (1 - m^2)(1 + 2m)$$

$$\phi_2(m) = 5(1 + m^2)$$

$$\phi_1(m) = 1 + m$$

Now, $\phi_3(m) = 0 \Rightarrow (1 - m^2)(1 + 2m) = 0$

$$\Rightarrow m = \pm 1, \frac{-1}{2}$$

$$m = 1,1,1-1.$$

Here, $m = \pm 1, \frac{-1}{2}$ we have non-repeated roots,

Hence, $c = -\dfrac{\phi_2(m)}{\phi_3'(m)}$

Now, $m = 1 \Rightarrow c = \dfrac{5}{3} \Rightarrow y = \dfrac{x+5}{3}$

$3y - x - 5 = 0$ is asymptote.

For $m = -1 \Rightarrow c = \dfrac{-5}{3} \Rightarrow y = \dfrac{-3x-5}{3}$

$3y + 3x + 5 = 0$ is 2nd asymptote.

and for $m = -\frac{1}{2} \Rightarrow c = -\frac{25}{6} \Rightarrow y = \frac{-3x-25}{6}$

$$6y+3x+25=0 \text{ is 3rd asymptote.}$$

EXAMPLE 7.16 Find asymptotes of the curve $(x^2y^2) = a^2(x^2 + y^2)$ which are parallel to the coordinates.

Ans. Given $(x^2y^2) = a^2(x^2 + y^2)$

$\therefore$
$$(x^2y^2) = a^2x^2 + a^2y^2$$

$\therefore$
$$(x^2y^2) - a^2x^2 = a^2y^2 \Rightarrow x^2(y^2 - a^2) = a^2y^2$$

For finding an asymptote parallel to the x-axis, the coefficient of $x = 0$,

$\therefore$
$$(y^2 - a^2) = 0 \Rightarrow y = \pm a$$

Given $(x^2y^2) - a^2x^2 = a^2y^2 \Rightarrow y^2(x^2 - a^2) = a^2x^2$.

For finding an asymptote parallel to the y-axis, the coefficient of $y = 0$,

$\therefore$
$$(x^2 - a^2) = 0 \Rightarrow x = \pm a$$

7.6 ASYMPTOTES OF nth DEGREE CURVATURE

Method 1: Let $f(x, y) = 0$ be n^{th} degree curve. Let $y - m_1x$ be a non-repeated linear factor of the nth-degree terms of the curve. One can also write

$$(y - m_1x) f_{n-1} + P_{n-1} = 0$$

where, f_{n-1} contains terms of degree n–1 and P_{n-1} contains other terms but less than $n-1$.

The curve has parallel asymptotes to line $(y - m_1x)$.

From the above equation, $(y - m_1x) = -\frac{P_{n-1}}{f_{n-1}}$

Now, taking the limit $x \to \infty, \frac{y}{x} \to m_1$

$$(y - m_1x) = \lim_{x \to \infty, \frac{y}{x} \to m_1} -\frac{P_{n-1}}{f_{n-1}}$$

By solving this equation, we will have asymptotes of a given curve.

Method 2: Let $f(x, y) = 0$ be n^{th} degree curve. Let $y - m_1x$ be a non-repeated linear factor of the nth degree terms of the curve and equation of the curve be,

$$(y - m_1x)^2 f_{n-2} + (y - m_1x)g_{n-2} + P_{n-2} = 0$$

where, f_{n-2} and g_{n-2} contains terms of degree $n-2$ and P_{n-2} contains other terms but less than $n-2$.

The curve has parallel asymptotes to line $(y - m_1x)^2$.

Further, divide by f_{n-2} and taking limit $x \to \infty, \frac{y}{x} \to m_1$

$$(y - m_1 x)^2 + (y - m_1 x) \lim_{\substack{x \to \infty \\ y/x \to m_1}} \frac{g_{n-2}}{f_{n-2}} + \lim_{\substack{x \to \infty \\ y/x \to m_1}} \frac{P_{n-2}}{f_{n-2}} = 0$$

$$\therefore \qquad (y - m_1 x)^2 + \lambda(y - m_1 x) + \mu = 0$$

Here, λ and μ are constants which is solved and asymptotes can be found.

EXAMPLE 7.17 Find asymptotes of the curve $(x^2 - y^2)(x + 2y) = y^2 - y + 1$.

Ans. Given $(x^2 - y^2)(x + 2y) = y^2 - y + 1$

$$\therefore \qquad (x - y)(x + y)(x + 2y) = y^2 - y + 1$$

Now, we have to find 3 parallel asymptotes to lines $(x - y) = 0$, $(x + y) = 0$, $(x + 2y) = 0$.

For the line, $(x - y) = 0$,

$$(x - y) = \frac{y^2 - y + 1}{(x + y)(x + 2y)}$$

Taking limit $x \to \infty, \frac{y}{x} \to 1 \left(Taking \ x - y = 0 \Rightarrow \frac{y}{x} = 1 \right)$

$$(x - y) = \lim_{\substack{x \to \infty \\ y/x \to 1}} \left(\frac{y^2 - y + 1}{(x + y)(x + 2y)} \right)$$

$$= \lim_{\substack{x \to \infty \\ y/x \to 1}} \left(\frac{\left(\frac{y}{x}\right)^2 - \frac{y}{x^2} + \frac{1}{x^2}}{\frac{(x + y)}{x} \frac{(x + 2y)}{x}} \right)$$

$$= \lim_{\substack{x \to \infty \\ y/x \to 1}} \left(\frac{1}{(1 + 1)(1 + 2)} \right)$$

$$(x - y) = \frac{1}{6}$$

Hence, $6x - 6y = 1$ is 1st asymptotes.

Now, 2nd asymptote parallel to $x + y = 0$,

$$(x+y) = \frac{y^2 - y + 1}{(x-y)(x+2y)}$$

Again, taking limits, $x \to \infty, \frac{y}{x} \to -1 \left(\textit{Taking } x + y = 0 \Rightarrow \frac{y}{x} = -1 \right)$

$$(x+y) = \lim_{\substack{x \to \infty \\ y/x \to -1}} \left(\frac{y^2 - y + 1}{(x-y)(x+2y)} \right)$$

Dividing by x^2 and applying the limits we have,

$$(x+y) = -\frac{1}{2}$$

Hence, $2x + 2y + 1 = 0$ will be 2nd asymptote.

A similar procedure for 3rd asymptote parallel to $x + 2y = 0$, $(x+2y) = \frac{y^2 - y + 1}{(x-y)(x+y)}$.

Limits will be, $x \to \infty, \frac{y}{x} \to -\frac{1}{2} \left(\textit{Taking } x + 2y = 0 \Rightarrow \frac{y}{x} = -\frac{1}{2} \right)$

$$(x+2y) = \lim_{\substack{x \to \infty \\ y/x \to -1}} \left(\frac{y^2 - y + 1}{(x-y)(x+y)} \right) = \frac{1}{3}$$

Therefore, $3x + 6y - 1 = 0$ will be 3rd asymptote.

EXAMPLE 7.18 Find asymptotes of the curve $(x - y - 1)^2(x^2 + y^2 + 2) + 6(x - y - 1)(xy + 7) - 8x^2 - 2x - 1 = 0$.

Ans. The factor $(x^2 + y^2 + 2)$ will give imaginary asymptotes

Asymptotes parallel to $(x - y - 1)^2$,

$$(x-y-1)^2 = -6(x-y-1) \lim_{\substack{x \to \infty \\ y/x \to 1}} \frac{(xy+7)}{x^2+y^2+2} - \lim_{\substack{x \to \infty \\ y/x \to 1}} \frac{-8x^2 - 2x - 1}{x^2 + y^2 + 2} = 0$$

$$\therefore \quad (x-y-1)^2 = -6(x-y-1) \lim_{\substack{x \to \infty \\ y/x \to 1}} \frac{\left(\frac{xy+7}{x^2}\right)}{\frac{x^2+y^2+2}{x^2}} - \lim_{\substack{x \to \infty \\ y/x \to 1}} \frac{\frac{-8x^2 - 2x - 1}{x^2}}{\frac{x^2 + y^2 + 2}{x^2}} = 0$$

$$\therefore \quad (x-y-1)^2 = -6(x-y-1)\frac{1}{2} + 4 = 0$$

Asymptotes are, $(x - y - 1) = 1$ and $(x - y - 1) = -4$

$$(x - y - 2) = 0 \text{ and } (x - y + 3) = 0.$$

7.7 INTERSECTION OF A CURVE AND ITS ASYMPTOTES

Generally, any straight line intersects n^{th} degree curve to n points but the asymptote is also a straight line, it touches the curve at some infinity point. Consequently, two coincident points of intersection will be at infinity.

Any asymptotes of nth degree curve in $n(n–2)$ points. The point of intersection of the curve and asymptotes lies on the curve, $F = F_1 - F_2$.

where, $F_1 = n^{th}$ degree equation

$F_2 =$ joint equation of n asymptotes

EXAMPLE 7.19 Shows that asymptotes of the curve $x^2y - xy^2 + xy + y^2 + x - y = 0$ cuts three points to its.

Ans. First we find asymptotes,

Parallel to the x-axis: $y = 0$

Parallel to the y-axis: $-x + 1 = 0 \Rightarrow x = 1$

$$\phi_3(m) = (x^2y - xy^2) = 0$$

$$\phi_2(m) = (xy + y^2)$$

Put $x = 1$, $y = m$ in terms,

$$\phi_3(m) = m - m^2 = 0 \Rightarrow m = 0, 1 \text{ and } \phi_3'(m) = 1 - 2m \text{ and } \phi_2(m) = m + m^2$$

Now, $c = -\dfrac{\phi_2(m)}{\phi_3'(m)} = -\dfrac{m + m^2}{1 - 2m}.$

at $m = 1$, $c = 2 \Rightarrow y = x + 2$.

So, there are 3 asymptotes:

Point of intersection $= n(n - 2) = 3(3 - 2) = 3$

Joint equation of asymptotes: $y(x - 1)(x - y + 2) = 0$

$$\Rightarrow x^2y - xy^2 + xy + y^2 - 2y = 0$$

and given equation: $x^2y - xy^2 + xy + y^2 + x - y = 0$.

Therefore, the point of intersection of the curve and asymptotes lies on the curve,

$$F = F_1 - F_2 \Rightarrow -2y - x + y = 0$$

$\therefore$

$$x + y = 0$$

7.8 ASYMPTOTES OF POLAR CO-ORDINATES

In polar coordinates, asymptotes can appear for curves that have a specific behavior as the radius r approaches infinity, or as the angle θ approaches certain values. These asymptotes describe the direction or path that the curve follows as it extends infinitely.

Let $f(\theta) = \frac{1}{r}$ be a polar curve. If θ_1 is one root of the equation $f(\theta) = 0$ then $r \sin(\theta - \theta_1) = \frac{1}{f'(\theta_1)}$ became asymptote.

Working Rule:

1. The equation of the form $f(\theta) = 1/r$.
2. Put $f(\theta) = 0$ and take $\theta = \theta_1$.
3. Find $f'(\theta)$ at $\theta = \theta_1$.
4. The asymptote will be, $r(\theta - \theta_1) = \frac{1}{f'(\theta_1)}$.

EXAMPLE 7.20 Find asymptotes of the curve $r \sin \theta = a \cos 2\theta$.

Ans. First convert into $f(\theta) = \frac{1}{r}$ form.

$$\frac{\sin \theta}{a \cos 2\theta} = \frac{1}{r}$$

Now, $f(\theta) = 0 \Rightarrow \frac{\sin \theta}{a \cos 2\theta} = 0$

$$\Rightarrow \sin\theta = 0$$

$$\Rightarrow \theta = n\pi$$

Next, $f'(\theta) = \frac{2a \sin 2\theta \sin \theta - a \cos 2\theta \cos \theta}{(a \cos 2\theta)^2}$

$$\therefore \quad f'(n\pi) = \frac{a \cos 2n\pi \cos n\pi - 2a \sin 2n\pi \sin n\pi}{(a \cos 2n\pi)^2}$$

$$\therefore \quad f'(n\pi) = \frac{(-1)^n}{a}$$

Equation of asymptote: $r \sin(\theta - n\pi) = \frac{a}{(-1)^n}$

$$\therefore \quad (-1)^n(-1)r \sin (n\pi - \theta) = a$$

$$\therefore \qquad (-1)^n(-1)r(-1)^{n-1}\sin(\theta) = a$$

$$\therefore \qquad r = \sin(\theta) = a$$

EXAMPLE 7.21 Find asymptotes of the curve $r(1 - e^{\theta}) = a$.

Ans. $\dfrac{(1-e^{\theta})}{a} = \dfrac{1}{r}$

Now,
$$f(\theta) = 0 \Rightarrow \frac{(1-e^{\theta})}{a} = 0$$
$$\Rightarrow (1-e^{\theta}) = 0$$
$$\Rightarrow e^{\theta} = 1$$
$$\Rightarrow \theta = 0 = \theta_1$$

Next,
$$f'(\theta) = \frac{-e^{\theta}}{a}$$

and
$$f'(0) = \frac{-e^{0}}{a} = \frac{-1}{a}$$

Equation of asymptote: $r\sin(\theta - 0) = -a$

$$r\sin\theta + a = 0$$

EXERCISES

Evaluate given problems:

1. Find the radius of curvature for the Cartesian curve $y = c\cosh\left(\dfrac{x}{c}\right)$.
2. Find the radius of curvature for the Cartesian curve $x = a(\cos\theta + \theta\sin\theta)$ and $y = a(\sin\theta - \theta\cos\theta)$.
3. Find the curvature of $x = 4\cos t$, $y = 3\sin t$. At what points ellipse does the curvature have maximum and minimum values?
4. Find the points on the parabola $y^2 = 8x$ at which the radius of curvature is 125/16.
5. For the curve $y = \dfrac{nx}{n+x}$, prove that $\left(\dfrac{2\rho}{n}\right)^{2/3} = \left(\dfrac{y}{x}\right)^2 + \left(\dfrac{x}{y}\right)^2$, where ρ is the radius of curvature.
6. Find the radius of curvature of $p = \dfrac{r^{m+1}}{a^m}$.

7. Find the radius of curvature of $p^2 = ar$.
8. Find the radius of curvature of the curve $p = a(1 + \sin \psi)$.
9. Find the radius of curvature of the curve $p \sin \psi = a$.
10. Find the radius of curvature at the origin of the curve $2x^3 - 3x^2y + 4y^3 + y^2 - 3x = 0$.
11. Find the circle of curvature of the curve $\sqrt{x}+\sqrt{y}=\sqrt{a}$ at a point $\left(\frac{a}{4}, \frac{a}{4}\right)$.
12. Find the circle of curvature of the curve $y^2 = 12x$ at a point (3, 6).
13. Find the chord of curvature through the pole $r = ae^{\theta \cot \alpha}$.
14. Find the asymptotes of the curve $x^2y + xy^2 - xy + y^2 + 3x = 0$.
15. Find the asymptotes of the curve $x^3 + 4x^2y + 5xy^2 + 2y^2 + 2x^2 + 4xy + 2y^2 - x - 9y + 1 = 0$.
16. Find asymptotes of the curve $(x^2y^2) - a^2(x^2 + y^2) - a^3(x + y) + a^4 = 0$ which are parallel to the coordinates.
17. Find asymptotes of the curve $x^2y^3 + x^3y^2 = x^3 + y^3$ which are parallel to the coordinates.
18. Find asymptotes of the curve $x^3 + 2x^2y + xy^2 - x^2 - xy + 2 = 0$ which are parallel to the coordinates.
19. Find asymptotes of the curve $(y - x)(y + x)(y - 2x) + y^2 - 6xy + 5x^2 - 2y + 2x + 3 = 0$.
20. Find asymptotes of the curve $(x^2 - 3x + 2)(x + y - 2) = 0$.
21. Find asymptotes of the curve $(x^2 - y^2)(x + 2y) + 5(x^2 + y^2) + x + y = 0$.
22. Find asymptotes of the curve $(x - 3y)^2(x^2 + 3y^2) - 4(x - 3y)xy - 3x^2 + xy + 2y = 0$.
23. Show that the eight points of intersection of the given curve and its asymptotes lie on the rectangular hyperbola. $x^4 - 5x^2y^2 + 4y^4 + x^2 - y^2 + x + y + 1 = 0$.
24. Show that the eight points of intersection of the given curve and its four asymptotes lie on the unit circle. $(x^2 - y^2)(y^2 - 4x^2) + 6x^3 - 5x^2y - 3xy^2 + 2y^3 - x^2 + 3xy - 1 = 0$.
25. Find asymptotes of the curve $r = a \sec \theta + b \tan \theta$.
26. Find asymptotes of the curve $r = a \log \theta$.
27. Find asymptotes of the curve $r(\pi + \theta) = ae^{\theta}$.

Answers:

1. y^2/c
2. at
3. Maximum at (4.0) is 4/9 and minimum at (0, 3) is 3/16.
4. $\left(\frac{9}{8}, \pm 3\right)$
5. Prove
6. $\rho = \frac{a^m}{(m+1)r^{m-1}}$.
7. $\rho = \frac{2r^{3/2}}{\sqrt{a}}$
8. $\rho = a$
9. $\rho = \frac{2p^3}{a^2}$
10. $\rho = \frac{3}{2}$
11. $\left(x - \frac{3a}{4}\right)^2 + \left(y - \frac{3a}{4}\right)^2 = \frac{a^2}{2}$
12. $(x-15)^2 + (y+6)^2 = 288$
13. $2r$
14. $y = 0, y + x = 0$
15. $2y + x + 2 = 0, y + x \pm \sqrt{2} = 0$
16. $x = \pm a, y = \pm a$
17. $x = \pm 1, y = \pm 1$
18. No asymptote parallel to the x-axis, parallel to the y-axis, $x = 0$
19. $y = x$
    $y + x + 2 = 0$
    $y - 2x = 1$
20. $x = 1$
    $x = 2$
    $x + y = 2$
21. $x + y - 5 = 0, 3(x - y) + 5 = 0,$
    $3(x + 2y) + 25 = 0$
22. $x - 3y = 2$
    $x - 2y = -1$
23. *Hint:* Equation of hyperbola: $x^2 - y^2 + x + y + 1 = 0$
25. $(-1)^n r \cos\theta = a + b$
26. $\theta = 0$
27. $r \sin\theta + ae^{-n} = 0$

8 Tracing of Curve and its Characteristics

8.1 TRACING OF CURVE

For any function, diametric representation defines the actual scenario of a function more easily and tells more than a description. A graph which is an image of that function gives the relationship between two quantities and helps predict a value corresponding to each other.

Just define a map $f: \mathrm{D} \to \mathrm{R}$ having points $\{(x, f(x) : x \in D)\}$. Here, we will plot the points. (x, y) which satisfies the equation $f(x, y) = 0$. This equation gives the functional relationship between given two variables. In other words, curve sketching or tracing is nothing but a plotting that satisfies functional relation. Further, we will see the curve tracing method for given Cartesian, parametric, and polar forms.

8.2 TRACING OF CARTESIAN CURVES

Plotting can be easy if the curve can be explicitly expressed as one function of another variable for any curve. Suppose the equation of a curve $f(x, y) = 0$ or $y = f(x)$ in Cartesian form. First, we find a region of the given curve. For example, if we take a curve $y = \tan^{-1}x$ lies in the first and third quadrants. Search out that the given curve is symmetric about any point or line.

(i) **Symmetric about the x-axis:** In the curve, if all the power of y is even then the curve is symmetric about the x-axis. And also $f(x, y) = f(x, -y)$. Here, we only have to draw one side of the curve at the x-axis and then its reflection will complete another portion of the curve.

(ii) **Symmetric about y-axis:** In the curve, if all the power of x is even then the curve is symmetric about the y-axis. And also $f(x, y) = f(-x, y)$. A similar process can be done we have performed in the above symmetricity to complete the graph of the curve.

(iii) **Symmetric about origin:** In the curve, if $f(x, y) = f(-x, -y) = 0$ even then curve is symmetric about the origin. In this case, one can draw the graph above the x-axis or y-axis and rotate at 180° will give you another part to complete it.

(iv) **Symmetric about line $y = x$:** If the equation of the curve does not change when we interchange x and y, then the given curve is symmetric about the line $y = x$.

(v) **Symmetric about line $y = -x$:** If the equation of the curve is unaltered when x and y are replaced by $-y$ and $-x$ respectively, then the given curve is symmetric about the line $y = x$.

(vi) **Symmetric in opposite quadrant:** If the equation of the curve remains unchanged when x and y are replaced by $-x$ and $-y$ respectively, then the given curve is symmetric in the opposite quadrant.

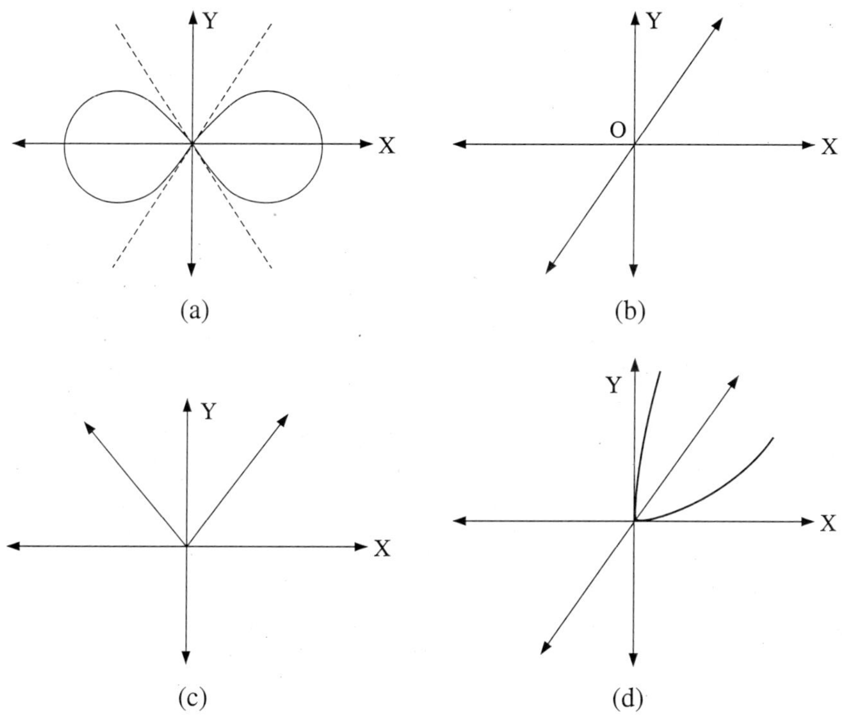

FIGURE 8.1 (a) Symmetric about x-axis; (b) Symmetric about origin; (c) Symmetric about y-axis; (d) Symmetric about line $y = x$.

After symmetricity is decided, obtain the points of intersection with the x-axis or y-axis by putting $y = 0$ or $x = 0$ respectively in the given equation of the curve. Then try to find the boundaries of the curve. Next, if the curve passes through the origin and tangent of the curve exists is obtained by equating lowest degree terms to zero. Finally, find the asymptotes to the curve parallel to the coordinate axes.

8.3 TRACING OF POLAR CURVES

Let the equation of the curve $f(x, y) = 0$ or $y = f(x)$ in Cartesian form then its polar form is given by $r = f(\theta)$. If the equation of the curve remains unaltered by replacing θ by $-\theta$, then the given polar curve is symmetric about the initial line. Then try to find the radius value for varying the value of θ from range 0 to π or 2π.

EXAMPLE 8.1 Draw the graph of the curve $x^{2/3} + y^{2/3} = a^{2/3}$.

Ans. The parametric equation of the curve $x = a\cos^3\theta$ and $y = a\sin^3\theta$.

$$(x^{1/3})^2 + (y^{1/3})^2 = a^{2/3}$$

This curve is known as Astroid. Since, if we replace x and y with $-x$ and $-y$, the equation remains unchanged, the curve is symmetric about the x and y axes. Its x-intercepts and y-intercepts are $(a, 0)$ and $(0, -a)$ respectively.

Now, by varying the value of θ we have x and y:

At $\theta = 0$, $(x, y) = (a, 0)$ and $\theta = \pi/2$, $(x, y) = (0, a)$

$$\theta = \pi/6,\ (x, y) = \left(\frac{3\sqrt{3}}{8}a,\ \frac{1}{8}a\right)$$

$$\theta = \pi/3,\ (x, y) = \left(\frac{1}{8}a,\ \frac{1}{2\sqrt{2}}a\right)$$

After calculating these values rough tracing can be done as follows:

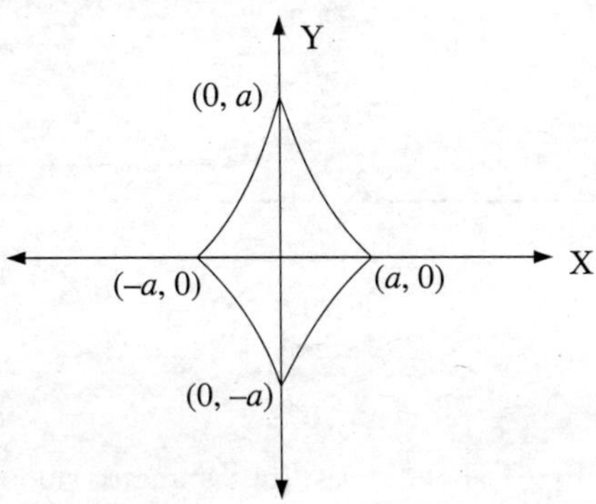

FIGURE 8.2 Tracing of Astroid.

EXAMPLE 8.2 Draw the graph of the curve $r = a(1 - \cos\theta)$.

Ans. Since the given function is even, the equation is unaltered and the curve is symmetric about the initial line. The curve is normally known as Cardioid.

Now, by varying the value of θ we have r:

At $\theta = 0, \quad r = 0$

$\theta = \pi/2, \quad r = a$

$\theta = \pi/6, \; r = a\left(1 - \dfrac{\sqrt{3}}{2}\right)$

$\theta = \pi/3, \; r = \dfrac{a}{2}$

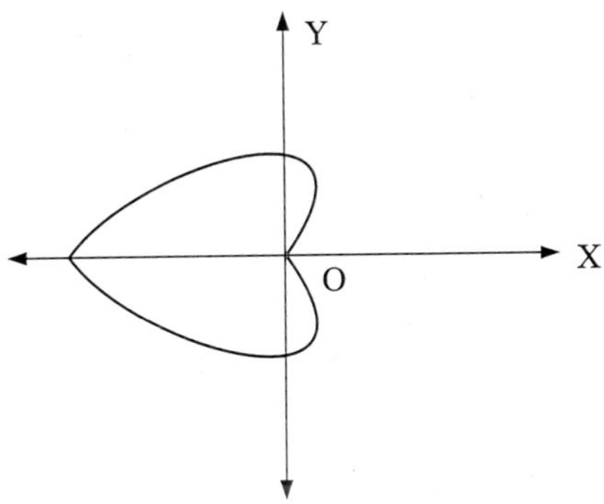

FIGURE 8.3 Tracing of Cardioid.

Singular point plays a vital role in algebraic curves. Curves are not smooth at these points. So, we have to analyze these points and their existence.

The other things for any tracing of a curve are concavity, convexity, and singular points are important things in calculus and mathematical analysis. The concavity and convexity of the graph of a function are signified by the change in the rate of slope of that function. It helps us to identify the curving of any graph or function while the singular point determines whether the curve transforms from concave upward to downward or convex upward means where concavity changes.

8.4 CONCAVITY AND CONVEXITY WITH RESPECT TO LINE

Let $f(x)$ and $g(x)$ be functions which is differentiable on an open interval I. If we take any curve and take a tangent to any point of the curve, if curve $g(x)$ lies on the upper part of the tangent at any point of the curve, then it is called concavity. Similarly, if curve $f(x)$ lies on the lower part of the tangent at any point of the curve, then it is called convexity.

There are also subtypes of concavity and convexity. The curve is said to be concave upward or convex downward at some point if the whole curve lies above the tangent at any point of the curve. In another sense, if the function is increasing at a specified point on a given interval. The curve is said to be

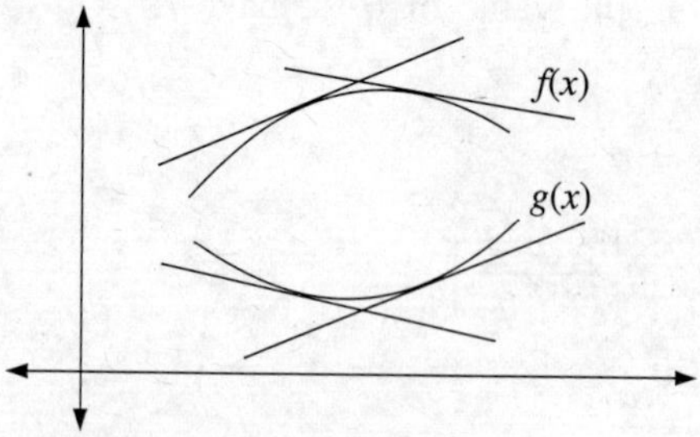

FIGURE 8.4 Concavity and convexity.

concave downward or convex upward if tangent at any point of the curve lies above the whole curve means the function will be decreasing at that point.

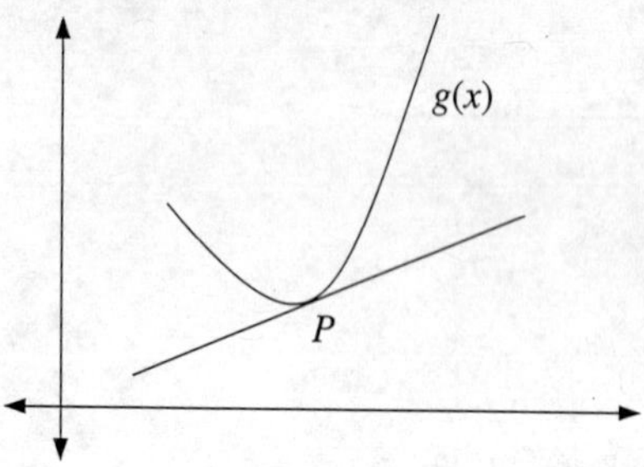

FIGURE 8.5 (a) Concave upward.

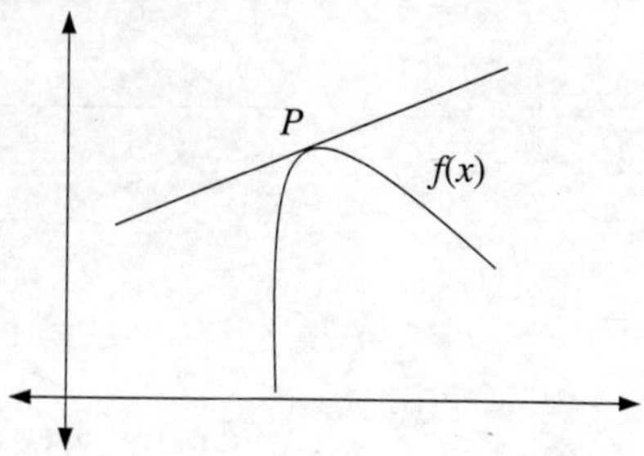

FIGURE 8.5 (b) Convex upward.

8.5 CONDITIONS FOR CONCAVITY AND CONVEXITY

Recall the thing that we have already studied in the previous section about the functions that, identify stationary points for any curve $y = f(x)$, we equate the first derivative to zero, i.e. $f'(x) = 0$. Further, for the classification of whether the given point is maximum or minimum, we evaluate the second derivative of a given function. Means if $f''(x) < 0$ at evaluated stationary point x, then point has maximum value and point has minimum value if $f''(x) > 0$ at x in a given function.

The points at which we are finding these optimum values need not be stationary points. If at any point in the function, the second derivative is negative $f''(x) < 0$ then we arrive arriving the condition for concavity implies the curve is **concave** at that point. Contrariwise, if the second derivative $f''(x) > 0$ is positive at any point, then the curve is **convex** at that point. One can interpret further that for concavity one interval of points lies around the maximum value and the same for convexity lies around the minimum value.

So, we come out with these conditions:

If f is a differentiable function on an interval I then, the graph of f is called,

(i) concave upward on *I*, $f'(x)$ is increasing function on the given interval. Condition for that, $f''(x) \geq 0$. If $f''(x) > 0$ then the graph is strictly concave upward.

(ii) Concave downward on *I*, $f'(x)$ is decreasing function on the given interval. Condition for that, $f''(x) \leq 0$. If $f''(x) < 0$ then the graph is strictly concave downward.

8.6 CONCAVITY AND CONVEXITY WITH RESPECT TO POINT

As we know concavity is the rate of change in the slope of any curve and the slope of a curve changes its sign represents the inflection point.

Second Derivative Test: Suppose $y = f(x)$ be any curve then $f''(x)$ exists on an open interval containing one point $x = c$ such that $f'(c) = 0$. Then

- If $f''(c) < 0$ then f has a relative maximum at $x = c$.
- If $f''(c) > 0$ then f has a relative minimum at $x = c$.
- If $f''(c) = 0$ or $f''(c)$ does not exist, the test is inconclusive and f may have relative minimum/ maximum or no extremum at all at $x = c$.

EXAMPLE 8.3 Use second derivative test to find relative extrema for $f(x) = \frac{1}{30}x^6 - \frac{1}{12}x^4$.

Ans. Given $f(x) = \frac{1}{30}x^6 - \frac{1}{12}x^4$

$$f'(x) = \frac{1}{5}x^5 - \frac{1}{3}x^3 = \frac{x^3}{5}\left(x^2 - \frac{5}{3}\right)$$

Now, $f'(x) = 0 \Rightarrow x = 0, \pm\sqrt{\frac{5}{3}}$

$\left(-\infty, -\sqrt{\frac{5}{3}}\right)$, $f'' > 0 \Rightarrow$ The function has a relative minimum

$\left(-\sqrt{\frac{5}{3}}, 0\right)$, $f'' = 0 \Rightarrow$ It is inconclusive

$\left(\sqrt{\frac{5}{3}}, \infty\right)$, $f'' > 0 \Rightarrow$ The function has a relative minimum

8.7 SINGULAR POINT

Definition: A point on the curve at which the curve shows an extraordinary behavior is called a singular point.

The nature of singularities is divided into two types:

1. Points of Inflection
2. Multiple Points

Initially, we only discuss about characteristics and conditions for the inflection point. Next, each will be discussed for multiple points.

8.8 POINTS OF INFLECTION

If at any point P to a curvature changes its concavity from being up to down or vice versa then P is called point of inflection.

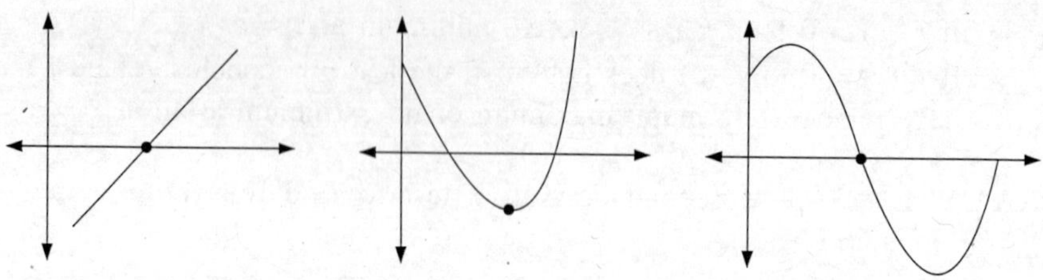

FIGURE 8.6 Point of Inflection.

While calculating the point of inflection for any graph of a function, the condition is that $f''(x)$ should exist and be continuous.

8.9 CONDITIONS FOR EXISTENCE OF A POINT OF INFLECTION

Method to find Interval of Concavity and Inflection points:

Step 1: Find all c for which $f''(c) = 0$ and solve or $f''(c)$ does not exist which divides the main domain of function into several intervals. Here, $f'''(x) \neq 0$ at point c.

Step 2: Now, for each interval calculate $f''(x)$. According to signs of $f''(x)$ as discussed above, decide the concavity of function.

Step 3: For each value of c in step 1, check the sign of $f''(x)$ to the left and right sides of the point $x = c$. Then changes in sign-on on two sides will give a point of inflection $(c, f(c))$.

EXAMPLE 8.4 Find the range of values of x for which the curve $y = x^4 - 6x^3 + 12x^2 + 5x + 7$ is curve has concavity.

Ans. Given $y = x^4 - 6x^3 + 12x^2 + 5x + 7$

$$\therefore \qquad \frac{dy}{dx} = 4x^3 - 18x^2 + 24x + 5$$

$$\therefore \qquad \frac{d^2y}{dx^2} = 12x^2 - 36x + 24$$

$$\therefore \qquad \frac{d^3y}{dx^3} = 24x - 36$$

Now, $f''(x) = 0 \Rightarrow 12x^2 - 36x + 24 = 0$

$$\therefore \qquad (x - 1)(x - 2) = 0$$

$f'''(x) \neq 0$ at $x = 1,2$.

Here, three intervals we have to consider

$(-\infty, 1)$, $\frac{d^2y}{dx^2} > 0 \Rightarrow$ the curve is concave upward

$(1,2)$, $\frac{d^2y}{dx^2} < 0 \Rightarrow$ the curve is concave downward

$(2, \infty)$, $\frac{d^2y}{dx^2} > 0 \Rightarrow$ the curve is concave upward

For point of inflection:

$$\frac{d^2y}{dx^2} = 0 \Rightarrow 12x^2 - 36x + 24 = 0$$

$\therefore x = 1,2$

At $x = 1 \Rightarrow y = 19$

And $x = 2 \Rightarrow y = 33$

Hence, the points of inflection are: (1, 19) and (2, 33).

EXAMPLE 8.5 Find the range of values of x for which the curve $y = 2x^6 - 5x^4 + 7x - 3$ is curve has concavity.

Ans. Given $y = 2x^6 - 5x^4 + 7x - 3$

$$\therefore \qquad \frac{dy}{dx} = 12x^5 - 20x^3 + 7$$

$$\therefore \qquad \frac{d^2y}{dx^2} = 60x^4 - 60x^2$$

Now, $$\frac{d^2y}{dx^2} = 0 \Rightarrow 60x^4 - 60x^2 = 0$$

Gives $x = 0, \pm 1$

$(-\infty, 0),\ \dfrac{d^2y}{dx^2} > 0 \Rightarrow$, the curve is concave upward

$(-1, 0),\ \dfrac{d^2y}{dx^2} < 0 \Rightarrow$, the curve is concave downward

$(0, 1),\ \dfrac{d^2y}{dx^2} < 0 \Rightarrow$, the curve is concave downward

$(1, \infty),\ \dfrac{d^2y}{dx^2} > 0 \Rightarrow$, the curve is concave upward

and the inflection points are: (–1,13) and (1,1).

EXAMPLE 8.6 Show that the point of inflection of the curve $y^2 = (x - a)^2 (x - b)$ lie on the straight line $3x + a = 4b$.

Ans. $y^2 = (x-a)^2(x-b) \Rightarrow y = (x-a)\sqrt{(x-b)}$

$$\therefore\quad \frac{dy}{dx} = \sqrt{(x-b)} + \frac{(x-a)}{2\sqrt{(x-b)}}$$

$$\therefore\quad \frac{dy}{dx} = \frac{2x-2b+x-a}{2\sqrt{(x-b)}} = \frac{3x-2b-a}{2\sqrt{(x-b)}}$$

$$\therefore\quad \frac{d^2y}{dx^2} = \frac{1}{2}\left(\frac{\sqrt{x-b}(3)-(3x-2b-a)\cdot\dfrac{1}{2\sqrt{x-b}}}{x-b}\right) = \frac{6(x-b)-(3x-2b-a)}{2(x-b)}$$

Now, $$\frac{d^2y}{dx^2} = 0 \Rightarrow \frac{6(x-b)-(3x-2b-a)}{2(x-b)} = 0$$

Implies $6(x-b)-(3x-2b-a)=0$

$$\therefore\quad (3x - 4b + a) = 0$$

$$\therefore\quad 3x + a = 4b$$

Also, at $x = \dfrac{4b-a}{3}$ we have $\dfrac{d^3y}{dx^3} \neq 0$.

Hence the proof.

EXAMPLE 8.7 If $y = ax^3 + 3bx^2$ has a point of inflection (–1, 2) then find the values of a and b.

Ans. Given $y = ax^3 + 3bx^2$

$$\therefore \qquad \frac{dy}{dx} = 3ax^2 + 6bx$$

$$\therefore \qquad \frac{d^2y}{dx^2} = 6ax + 6b \text{ and}$$

$$\therefore \qquad \frac{d^3y}{dx^3} = 6a$$

The point of inflection is (–1, 2).

$$2 = a(-1)^3 + 3b(-1)^2 \Rightarrow -a + 3b = 2$$

We have a condition that, $\dfrac{d^2y}{dx^2} = 0 \Rightarrow 6ax + 6b = 0$

As a point of inflection is (–1, 2) $\Rightarrow 6a(-1) + 6b = 0$

$$\Rightarrow a = b$$

This $a = b$ substituting in $-a + 3b = 2$ we have, $a = b = 1$

More, $\dfrac{d^3y}{dx^3} \neq 0 \rightarrow a \neq 0.$

8.10 INFLECTION POINTS FOR POLAR CURVES

Let $r = f(\theta)$ be an equation of a curve in polar coordinates (r, θ), where r is the radial vector and the radial vector makes an angle θ with the initial line.

A curve $r = f(\theta)$ is called concave to the pole if perpendicular distance at any point say $P(r, \theta)$ from the pole to tangent increases with an increase in r and if it decreases then it will be considered as convex.

A curve is concave to the pole if dp/dr is positive and convex to the pole if dp/dr is negative.

For point of inflection, $\dfrac{dp}{dr} = 0$

Take $r_1 = \dfrac{dr}{d\theta}$ and $r_2 = \dfrac{d^2r}{d\theta^2}$

The radius of curvature $= r\dfrac{dr}{dp} = \dfrac{(r^2 + r_1^2)^{3/2}}{r^2 + 2r_1^2 - rr_2}$

Now, $\dfrac{dp}{dr} = \dfrac{r^2 + 2r_1^2 - rr_2}{(r^2 + r_1^2)^{3/2}}$

Hence, the curve is concave or convex to the pole or origin according to $r^2 + 2r_1^2 - rr_2$ is negative/positive. For point of inflection $\frac{dp}{dr} = 0 \Rightarrow r^2 + 2r_1^2 - rr_2 = 0$.

Take $r = \frac{1}{v} \Rightarrow v = \frac{1}{r}$

$$\Rightarrow \frac{dv}{d\theta} = -\frac{1}{r^2}\frac{dr}{d\theta}$$

$$\Rightarrow \frac{d^2v}{d\theta^2} = -\frac{1}{r^2}\frac{d^2r}{d\theta^2} + \frac{2}{r^3}\left(\frac{dr}{d\theta}\right)^2$$

$$\Rightarrow r^3\frac{d^2v}{d\theta^2} = -r\frac{d^2r}{d\theta^2} + 2\left(\frac{dr}{d\theta}\right)^2$$

$$\Rightarrow \frac{1}{v^3}\frac{d^2v}{d\theta^2} = -rr_2 + 2r_1^2$$

We have, $r^2 + 2r_1^2 - rr_2 = r^2 + \frac{1}{v^3}\frac{d^2v}{d\theta^2} = 0$

$$\Rightarrow \frac{1}{v^2} + \frac{1}{v^3}\frac{d^2v}{d\theta^2} = 0$$

$$\Rightarrow v + \frac{d^2v}{d\theta^2} = 0$$

Thus, according to the sign of the above expression behavior of the curve is decided.

EXAMPLE 8.8 Find point of inflection for $r^2\theta = a^2$.

Ans. $r^2\theta = a^2 \Rightarrow r = \frac{a}{\sqrt{\theta}}$

$$\Rightarrow r = \frac{1}{v} = \frac{a}{\sqrt{\theta}}$$

$$\Rightarrow v = \frac{\sqrt{\theta}}{a}$$

$$\Rightarrow \frac{dv}{d\theta} = \frac{1}{2a\sqrt{\theta}}$$

$$\Rightarrow \frac{d^2v}{d\theta^2} = -\frac{1}{4a}\theta^{-3/2}$$

Condition for point of inflection is,

$$v + \frac{d^2v}{d\theta^2} = 0 \Rightarrow \frac{\sqrt{\theta}}{a} - \frac{1}{4a}\theta^{-3/2} = 0$$

$$\Rightarrow \frac{1}{a}(4\theta^2 - 1) = 0$$

$$\Rightarrow \theta = \pm\frac{1}{2}$$

$\theta = -\frac{1}{2}$ is not a valid point. For $\theta = \frac{1}{2}, r = \frac{a}{\sqrt{\theta}} = \sqrt{2}a$

So, $(r, \theta) = \left(\sqrt{2}a, \frac{1}{2}\right)$

8.11 MULTIPLE POINTS

A point on the curve at which more than one branch of the curve passes is called a Multiple Point.

The classification of multiple points includes double points, triple points, and higher-order multiple points. If a point on a curve through which two branches of the curve pass is called a Double Point, if three branches of a curve pass is called a Triple Point so on if p branches of a curve pass is called a Multiple Point of pth order.

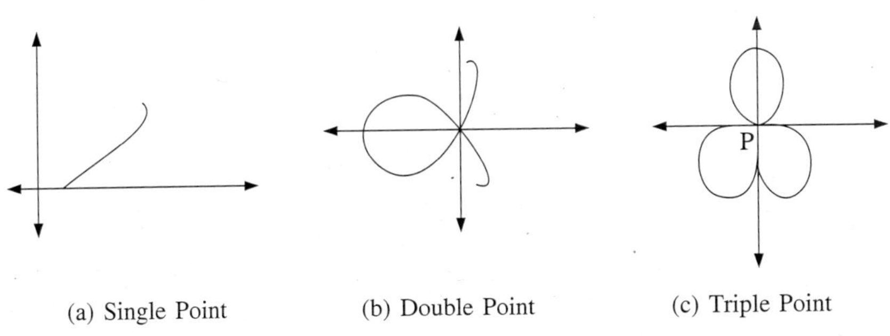

(a) Single Point (b) Double Point (c) Triple Point

FIGURE 8.7 Multiple Point.

8.12 CLASSIFICATION OF DOUBLE POINTS

Now, if two branches of a curve pass at a point on the curve then the point is called Double Point. This classification includes nodes, cusps, and isolated points.

(i) **Node:** A double point P on curve is called a Node if two real distinct branches pass through point P on curve having distinct tangents. In given Figure 8.8, C_1 and C_2 are branches, P is Node and T_1, T_2 are tangent at that point P.

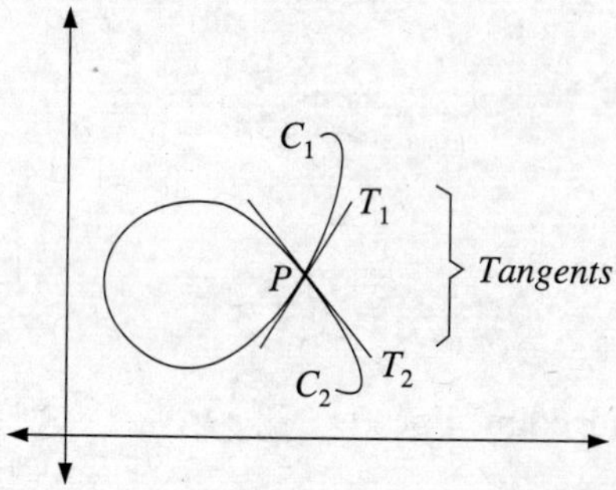

FIGURE 8.8 Node.

(ii) **Cusps:** A double point P is called a Cusp if two real branches of a curve pass through point P having common tangents. In other words, for two branches of a curve if we can draw a common tangent at a point of intersection then that double point is Cusp.

There is further classification in Cusps which are of two types. And each type contains species.

(a) ***Single Cusp:*** If the curve lies on the same sides as common normal then it is a Single cusp. In this single cusp, if branches of a curve lie on the same sides of the common tangent, then it will be named as a single cusp of the 1st species and it is a single cusp of 2nd species if branches curve lie on opposite sides of the common tangent. In Figure 8.6, P is the Cusp and T is tangent at that point P.

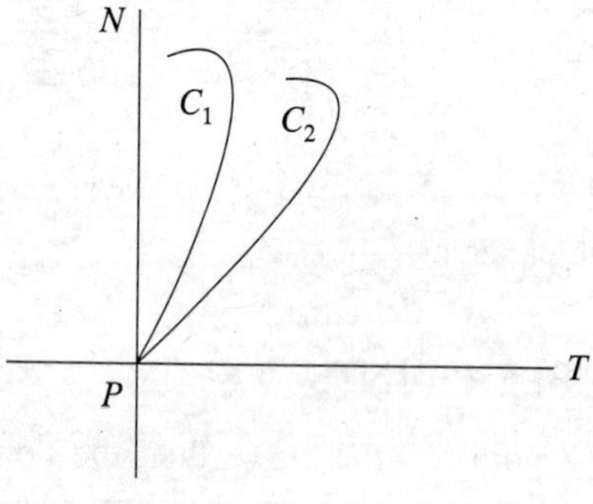

FIGURE 8.9 (a) Single Cusp of 1st species.

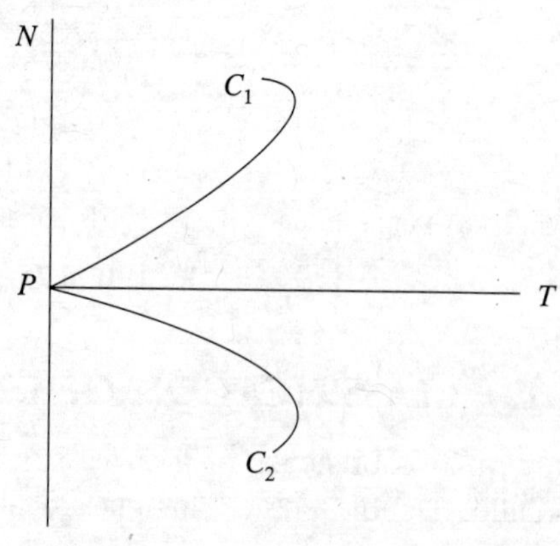

FIGURE 8.9 (b) Single Cusp of 2nd species.

(b) ***Double Cusp:*** If branches of a curve lie on both sides of the common normal then it is Double cusp. Precisely, if branches of a curve lie on both sides of normal and the same sides of common tangent then it will be named as double cusp of 1st species and it is double cusp of 2nd species if branches curve lie on opposite sides of common tangent and both sides of normal.

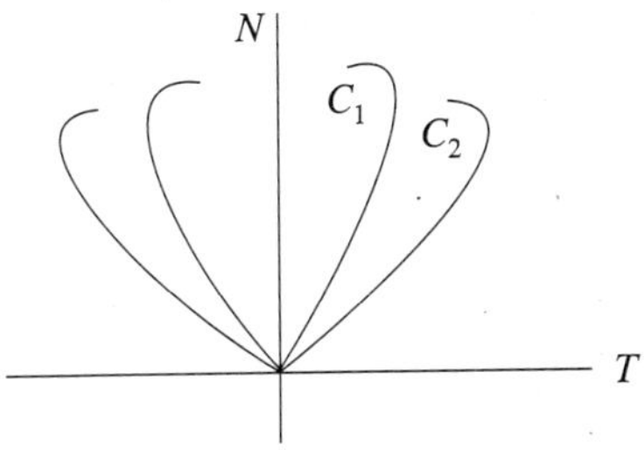

FIGURE 8.10 (a) Double Cusp of 1st species.

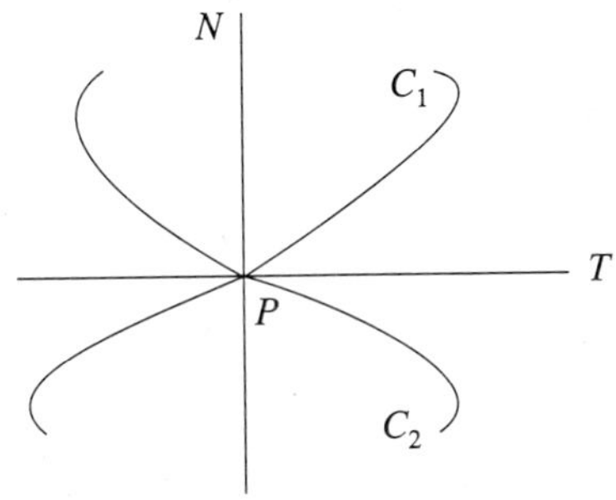

FIGURE 8.10 (b) Double Cusp of 2nd species.

Thus, we have a double point specifically the Cusp, and further if branches lie on the same side, then it is a single/double cusp of 1st kind, and so on.

(iii) **Isolated Point (Conjugate Point):** A point is called an isolated or conjugate point if there exist no other real points on the curve in its neighborhood.

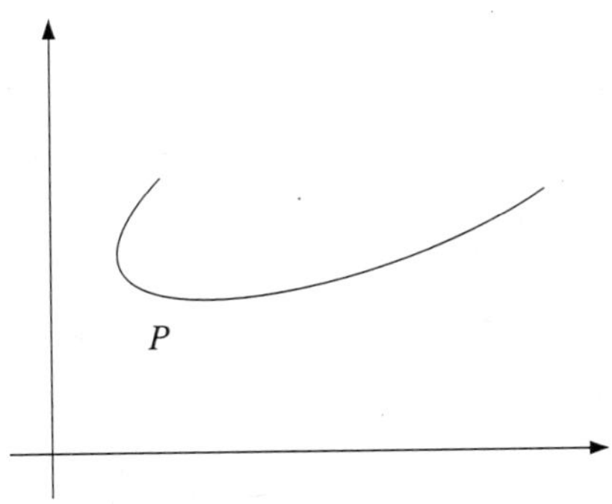

FIGURE 8.11 Isolated Point.

8.13 CONDITION FOR THE EXISTENCE OF DOUBLE POINTS

Let $f(x, y)$ be a curve. Assume a point $P(x_0, y_0)$ on a curve. $P(x_0, y_0)$ be a double point if at least two branches of the curve intersect at the point $P(x_0, y_0)$. If the equation of the curve is given by $f(x, y) = 0$, then to derive the condition for double point $P(x_0, y_0)$ exists we have steps,

1. Find $\frac{\partial f}{\partial x}$ and $\frac{\partial f}{\partial y}$.
2. Partial derivatives of $f(x, y)$ at $P(x_0, y_0)$ should vanish implies, $\left.\frac{\partial f}{\partial x}\right|_{(x_0, y_0)} = 0$ and $\left.\frac{\partial f}{\partial y}\right|_{(x_0, y_0)} = 0$
3. By solving these two, we have the values $x = (x_1, x_2, x_3, ...)$ and $y = (y_1, y_2, y_3, ...)$. So, a possible combination of points is $\{(x_1, y_1), (x_1, y_2), (x_1, y_3), (x_2, y_1), ...\}$.
4. Choose the point that satisfies the given curve equation $f(x, y) = 0$. Then find $\frac{\partial^2 f}{\partial x^2}$, $\frac{\partial^2 f}{\partial y^2}$ and $\frac{\partial^2 f}{\partial x \partial y}$ at those selected points.
5. Evaluate $\left(\frac{\partial^2 f}{\partial x \partial y}\right)^2 - \left(\frac{\partial^2 f}{\partial x^2}\frac{\partial^2 f}{\partial y^2}\right)$. Then if

   (a) $\left(\frac{\partial^2 f}{\partial x \partial y}\right)^2 - \left(\frac{\partial^2 f}{\partial x^2}\frac{\partial^2 f}{\partial y^2}\right) > 0$, the double point is a node.

   (b) $\left(\frac{\partial^2 f}{\partial x \partial y}\right)^2 - \left(\frac{\partial^2 f}{\partial x^2}\frac{\partial^2 f}{\partial y^2}\right) < 0$, the double point is conjugate.

   (c) $\left(\frac{\partial^2 f}{\partial x \partial y}\right)^2 - \left(\frac{\partial^2 f}{\partial x^2}\frac{\partial^2 f}{\partial y^2}\right) = 0$, the double point is the cusp.

These conditions state that the curve is not well defined and two branches are intersecting at a point which proves the point is double.

EXAMPLE 8.9 Find the position and nature of double points on the curve $x^4 + 4ax^3 - 2ay^3 + 4a^2x^2 + 3a^2y^2 - a^4 = 0$.

Ans. Given $f(x,y) = x^4 + 4ax^3 - 2ay^3 + 4a^2x^2 + 3a^2y^2 - a^4 = 0$

$$\frac{\partial f}{\partial x} = 4x^3 + 12ax^2 + 8a^2x \text{ and } \frac{\partial^2 f}{\partial x^2} = 12x^2 + 24ax + 8a^2$$

$$\frac{\partial f}{\partial y} = -6ay^2 + 6a^2y \text{ and } \frac{\partial^2 f}{\partial y^2} = -12ay + 6a^2$$

$$\frac{\partial^2 f}{\partial x \partial y} = 0$$

For double points:

$$\frac{\partial f}{\partial x} = 0 \Rightarrow 4x^3 + 12ax^2 + 8a^2x = 0$$

$$\Rightarrow 4x(x^2 + 3ax + 2a^2) = 0$$

$$\Rightarrow x = 0, \ -2a, \ -a$$

$$\frac{\partial f}{\partial y} = 0 \Rightarrow -6ay^2 + 6a^2y = 0$$

$$\Rightarrow -6ay(y - a) = 0$$

$$\Rightarrow y = 0, a$$

Possible points are: (0, 0), (0, a), ($-2a$, 0), ($-2a$, a), ($-a$, 0), ($-a$, a)

Out of these points (0, 0), ($-2a$, a) and ($-a$, 0) satisfy the equation $f(x,y) = x^4 + 4ax^3 - 2ay^3 + 4a^2x^2 + 3a^2y^2 - a^4 = 0$.

At (0, 0)

$$\left(\frac{\partial^2 f}{\partial x \partial y}\right)^2 - \left(\frac{\partial^2 f}{\partial x^2}\frac{\partial^2 f}{\partial y^2}\right) = 0 - (8a^2)(-6a^2) = 48a^2 > 0$$

(0, 0) is node.

Similarly, since at ($-2a$, a) and ($-a$, 0), $\left(\frac{\partial^2 f}{\partial x \partial y}\right)^2 - \left(\frac{\partial^2 f}{\partial x^2}\frac{\partial^2 f}{\partial y^2}\right) > 0$, points are nodes.

EXAMPLE 8.10 Find the position and nature of double points on the curve $x^3 + 2x^2 + 2xy - y^2 + 5x - 2y = 0$.

Ans. Given $f(x,y) = x^3 + 2x^2 + 2xy - y^2 + 5x - 2y = 0$

$$\frac{\partial f}{\partial x} = 3x^2 + 4x + 2y + 5 \text{ and } \frac{\partial^2 f}{\partial x^2} = 6x + 4$$

$$\frac{\partial f}{\partial y} = 2x - 2y - 2 \text{ and } \frac{\partial^2 f}{\partial y^2} = -2$$

$$\frac{\partial^2 f}{\partial x \partial y} = 2$$

For double points:

$$\frac{\partial f}{\partial x} = 0 \Rightarrow 3x^2 + 4x + 2y + 5 = 0$$

$$\frac{\partial f}{\partial y} = 0 \Rightarrow 2x - 2y - 2 = 0$$

$\Rightarrow x = y + 1$ put into the above equation we have,

$$\Rightarrow 3(y+1)^2 + 4(y+1) + 2y + 5 = 0$$

$$\Rightarrow 3y^2 + 12y + 12 = 0$$

$$\Rightarrow y = -2, -2$$

As $x = y + 1 \Rightarrow x = -1, -1$

Only one possible combination of points is: $(-1, -2)$

At where, $\left(\frac{\partial^2 f}{\partial x \partial y}\right)^2 - \left(\frac{\partial^2 f}{\partial x^2}\frac{\partial^2 f}{\partial y^2}\right) = 4 - 4 = 0$

$(-1, -2)$ is a cusp.

8.14 TANGENTS AT THE ORIGIN

The tangents at the origin are obtained by equating to zero to the lowest degree terms in the given equation of the curve then we have real or imaginary tangents.

The working rule for finding the nature of the Double point at the origin:

(i) If the resulted tangents at the origin are real and distinct then the origin is a node.

(ii) If the resulted tangents at the origin are real and coincident then the origin is a cusp.

(iii) If the resulted tangents at the origin are imaginary then the origin is an isolated/conjugate point.

If the tangents at the origin are $y^2 = 0$ then solve the equation of the curve for y. Later neglect all terms of x having powers more than 2. After this if small values $x \neq 0$, y is real then the origin is also real, otherwise imaginary. Accordingly, you can conclude the nature of the origin. This procedure can also be used when the tangents at the origin are $x^2 = 0$.

EXAMPLE 8.11 Find the nature of origin on the curve $y^2(a^2 + x^2) = x^2(a^2 - x^2)$.

Ans. Given $y^2a^2 + y^2x^2 - x^2a^2 + x^4 = 0$

Now equating to zero to the lowest degree terms,

$$a^2(y^2 - x^2) = 0 \Rightarrow y = \pm x$$

Tangents at the origin are real and distinct hence the tangent is a node.

EXAMPLE 8.12 Find the nature of origin on the curve $y^2 - 4x^2y + x^5 = 0$.

Ans. Given $y^2 - 4x^2y + x^5 = 0$

Equating to zero to the lowest degree terms,

$$y^2 = 0 \Rightarrow y = 0$$

The curve may have coincident tangents hence either origin is cusp or conjugate.

Since the given function is quadratic, $y^2 - 4x^2y + x^5 = 0$, we have roots,

$$y = \frac{4x^2 \pm \sqrt{16x^4 - 4x^5}}{2} = \frac{4x^2 \pm 2x^2\sqrt{4-x}}{2}$$

For the small value of $x \neq 0$, will be real then the origin is cusp.

EXERCISES

Evaluate given problems:

1. Find the range of values of x for which the curve $y = x^3 - 6x^2 + 9x + 30$ is curve has concavity.
2. Find the range of values of x for which the curve $y = x^{1/3} + 2$ is curve has concavity.
3. Find the intervals for which the curve $y = 3x^5 - 40x^3 + 3x - 20$ is curve has concavity.
4. Find the intervals for which the curve $y = x + \frac{4}{x}$ is curve has concavity.
5. Examine the point of inflection of the curve and concavity in $(0, 2\pi)$.
6. Show that the curve $x = \log\left(\frac{y}{x}\right)$ has a point of inflection at $(-2, -2e^{-2})$.
7. Find the values of a and b such that the curve $y = ax^3 + bx^2$ has a point of inflection at $(-1, 2)$.
8. Find relative extrema for $f(x) = x^5 - 5x^3$.
9. Find a point of inflection for $r(\theta^2 - 1) = a\theta^2$.
10. Find a point of inflection for $re^{\theta} = a(1 + \theta)$.
11. Determine the position and nature of the double point of the curve $x^3 + x^2 + y^2 - x - 4y + 3 = 0$.
12. Determine the position & nature of the double point of the curve $x^3 - y^2 - 7x^2 + 4y + 15x - 13 = 0$.

Determine the position and nature of the double point of the curve:

13. $x^2(x-2)^3 = y(y-6)+9$
14. $x^2(x-y)+y^2=0$
15. $x^4+y^3-2x^3+3y^2=0$
16. $2(x^3+y^3)-3(3x^2+y^2)+12x-4=0$

Find the nature of origin on the curve:

17. $x^3+y^3-3axy=0$
18. $x^4+y^3+2x^2+3y^2=0$
19. $(2x+y)^2-6xy(2x+y)-7x^3=0$

Answers

	Interval	*Concavity*	*Inflection point*
1.	$(-\infty, 2)$	concave down	(2, 32)
	$(2, \infty)$	concave up	
2.	$(-\infty, 0)$	concave up	(0, 2)
	$(0, \infty)$	concave down	
3.	$(-2, 0)$ and $(2, \infty)$	concave up	
	$(-\infty, -2)$ and $(0, 2)$	concave down	
4.	$(-\infty, 0)$	concave down	
	$(0, \infty)$	concave up	
5.	$(\pi, 2\pi)$	concave up	$x = n\pi$
	$(0, \pi)$	concave down	

6. proof 7. $a = 1$, $b = 3$

8. Local maximum at $x = -\sqrt{3}$, Local minimum at $x = \sqrt{3}$ 9. $\left(\frac{3a}{2}, \pm\sqrt{3}\right)$

10. No point of inflection 11. (–1,2) is node 12. (3,2) is node

13. (0,3) and (2,3) are isolated point 14. (0,0) is cusp 15. (0,0) is cusp

16. $x = y = 0$, node 17. $y = \pm i\sqrt{\frac{2}{3}}x$, conjugate 18. (0,0) is cusp

19. Cusp

Index